Springer Series in Optical Sciences   Volume 41
Edited by Jay M. Enoch

SEP/ae
PHYS

Springer Series in Optical Sciences

Editorial Board: J.M. Enoch D.L. MacAdam A.L. Schawlow K. Shimoda T. Tamir

Advances in Diagnostic Visual Optics

Proceedings of the Second International Symposium, Tucson, Arizona, October 23–25, 1982

Editors

G. M. Breinin and I. M. Siegel

With 197 Figures

Springer-Verlag
Berlin Heidelberg New York Tokyo 1983

69765947

PHYS

Professor Dr. GOODWIN M. BREININ
Professor Dr. IRWIN M. SIEGEL
Department of Ophthalmology, New York University Medical Center, 550 First Avenue
New York, NY 10016, USA

ISBN 3-540-13079-9 Springer-Verlag Berlin Heidelberg New York Tokyo
ISBN 0-387-13079-9 Springer-Verlag New York Heidelberg Berlin Tokyo

Foreword

RE76
A38
1983
PHYS

Opening Remarks of the President, 2nd ISVO,
Professor G.M. Breinin, M.D.

The study of visual processes is surely unique as a clinical specialty, incorporating the disciplines of physics, chemistry, physiology, and psychology. Diagnosing and correcting disorders of the visual system in these
last two decades of the 20th century has brought all of us into close proximity with computer sciences, laser technology, the marvels of electronic
microcircuitry, and the impressive developments in optical materials. During the course of this meeting we shall be hearing about how these different
technologies can interact with one another, and we shall discover that such
interaction may produce new diagnostic tools and new optical devices. We
shall also learn that the optical qualities of the eye change during life,
producing subtle and complex alterations in vision.

On behalf of the members and organizing committee of the American Committee on Optics and Visual Physiology and our co-sponsoring organization, the
Optical Society of America, I welcome you to this second symposium on visual
optics. The first symposium took place in Japan in 1978 and, like the present one, was a satellite meeting of the International Congress of Ophthalmology.

The third symposium in this series will take place in Italy in 1986 as
part of the next session of the International Congress of Ophthalmology.
I look forward to seeing all of the present participants at that time, as
we follow the continuing advancements in this fascinating, rapidly developing area of visual research.

So welcome to the new frontiers in vision which we celebrate here at the
frontier of America's old west.

Acknowledgements

We would like to thank the following organizations who lent generous support to the symposium:

Canon Corporation

Mentor Corporation

Nikon: Nippon Kogaku K.K.

Pilkington Brothers Ltd.

Rodenstock, GmbH

Society of Worshipful Spectacle Makers, U.K.

Sonometric Systems, Inc.

Topcon Instrument Corporation

UKO International Ltd.

Warner-Lambert: American Optical Scientific Instrument Group

The Optical Society of America co-sponsored the symposium and provided valuable organizational assistance.

Contents

Part 3 Applied Optics

Part 4 Electronic Visualization of the Fundus

Part 5 Developments in Visual Electrodiagnostic Techniques

Part 6 Clinical Applications of Visual Psychophysical Testing

Part 1

Optical Characteristics of the Eye

On the Anatomy of Ocular Optics

Robert A. Weale

Institute of Ophthalmology, University of London, Judd Street
London, WC1 9QS, United Kingdom

One of the many paradoxes that make the study of visual science so attractive is the examination of the relation between ocular structure and function: few anatomists have felt the need to master optics, and how many optical experts have concerned themselves with embryological or developmental detail is anyone's guess. Perhaps the number of paradoxes is only increased if someone unable to claim expertise in either field steps into the breach. There are, however, worse ways of idling away a few minutes, and who knows? but something of interest may emerge from the attempt.

The subject matters to which I wish to turn our attention are the following:
(1) Is there a ready explanation of physiological astigmatism?
(2) Why is there a tendency for myopia amongst those born prematurely?
(3) What does birth do for our eyes?
(4) Is there a difference between men and women?

1. Physiological Astigmatism

The overwhelming evidence is that we are born with corneae that are more sharply curved in the vertical than the horizontal meridian. The astigmatism is with the rule. MARIN-AMAT [1] attributed this tentatively to babies crying a great deal, constricting the orbicularis, and thereby so squeezing the eye as to deform the cornea whence it assumes the form observed in early life. During later life, when the medial recti pull on the cornea and flatten them in the horizontal meridian, with-the-rule is said to give way to against-the-rule, a view shown to be untenable by READING [2] who noted a similar change in congenitally blind persons in which convergence does not occur. I wish to suggest that physiological astigmatism is the result not of function but of growth.

Some sixty years ago, LESER [3] published a French description of ocular development. This has been ignored by the Anglo-Saxon literature because the latter does not specialize in French, or, at any rate, because the Madame Curie of ocular embryology, namely Dame Ida Mann, probably commanded it inadequately. What Leser showed is fascinating: the foetal eye is not ball-shaped. His photographs are hard to reproduce but the situation is clear from Fig. 1, which shows an embryo 10 mm in length. The oblate shape of the rudimentary eye is plain, with the major axis, the prospective optical axis, running through the lens rudiment. Fig. 2, which shows the embryo at an earlier stage (4.2 mm), suggests that the ellipsoid nature is the result of ectodermal growth inertia: the cells seem to grow "forward".

These observations help to explain not only why the sagittal length is greater than any of the other diameters of the eye-ball, but also why there

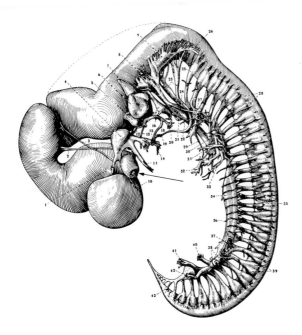

Fig. 1 10 mm long embryo
[24]. The arrow points at
the eye

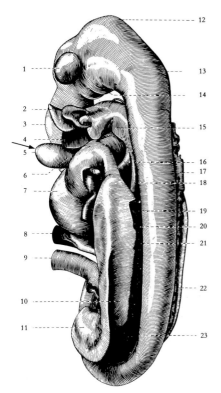

Fig. 2 4.2 mm long embryo [24]. The
arrow points at the optic vesicle

3

is an anlage for all the diameters to be dissimilar (Fig. 3). The matter,
however, does not rest there. For a start, if no further change occurred,
the optic disc would come to lie on the optic axis, hardly a situation calculated to achieve good visual results. At this stage, the eyes are still
lateral outcrops of the brain. In order to make possible forward and fused
binocular vision, they migrate from their lateral positions, a circumstance
that not only explains why they are attached to the rear of the now folded
brain but also why they appear to be rotated in relation to a brain that has
itself turned about an axis perpendicular to its principal growth line. This
movement is accompanied by rotation as is apparent when one studies the projection of the visual field on sections of the visual path at various prox-
imities.

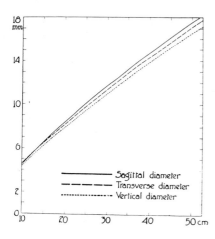

Fig. 3 The three principal dimaters of
the human eye-ball [24]

The final point to consider is the growth of the ocular ellipsoid. Once
the eyes have set out on their journey to the frontal part of the head, there
occurs a gross asymmetry in the rate of the growth of the nasal and temporal
moieties respectively. The latter grow appreciably faster. The result is
that the optic disc appears to move nasally, away from the optic axis of the
eye. Note that, in point of fact, it is better to view the situation in a
reverse sense: the temporal part of the eye, including the anterior segment
swings outward. Since in about 80% of people [4] the angle $\propto$ between the
visual and optic axes is positive, this can be viewed as an adaptational in-
sufficiency: the optic axis tries to catch up with the visual axis, but fails
to do so in the majority of eyes.

We can now understand a little more about the shape of the cornea. Given
that the eye is not a ball, there is no way in which a smooth spherical surface can be grafted onto it to form the cornea. In the new-born the cornea
and the eye-ball are part of the same continuous surface. After birth, however, the cornea may stretch a great deal but grows very little. Hence it
starts to protrude from the growing eye-ball. The above-mentioned embryonic
temporal growth does not lead to any asymmetric meridional change in the corneal radius of curvature and is, therefore, unlikely to cause the decrease
in horizontal curvature that manifests as astigmatism.

In fine, the non-sphericity of the young cornea is held to stem from that
of the eye-ball as a whole. As a result of the intra-ocular pressure there

are tensile forces set up in the cornea, the component in the horizontal meridian being greater than is true of the vertical. It follows that corneal birefringence [5], which is a consequence of the ensuing anisotropy is also explicable in terms of the oblate shape of the eye-ball.

2. Prematurity

This condition affects some 6% of live births in the West but shows a considerable variation across the world. Insofar as the eyes are concerned it presents a remarkable biological puzzle. As everyone knows, kittens are born blind, and can see only eight days post natum. There are close parallels between their retinal and receptoral development, physiological function, and behaviour. Other species, e.g. chicks, have virtually fully developed vision the moment they are born. But what is the explanation of the fact that, with the normal term for man being some 40 weeks, normal vision is attainable in premature individuals born, say, after 30 weeks? We cannot go into speculations on whether the long term is a relatively late evolutionary adaptation, accompanying the increase of the brain that distinguishes us from other primates. From an economic point of view, the premature readiness of the retina seems remarkably wasteful, though of evident benefit to the small but impatient minority.

What can be said of the optics of the premature eye? The foetal lens is more nearly spherical than that of adults and this would lead one to expect the normal new-born eye to be relatively myopic. In fact, the reverse is found in both Western and Japanese babies: hypermetropia is the rule of the day at birth. The explanation may well lie in the shape of the cornea. At birth, the sparse available data lead one to think there is a rapid increase in the radius of curvature (Fig. 4), probably owing to the antenatal growth of the eye-ball. Although only a few results for the premature radius of curvature have so far been obtained [6], it is clear that a small radius of curvature (Table 1) will predispose the eye to myopia (cf. Section 3). The tendency for a smaller anterior chamber in premature eyes as compared with normal ones will contribute to this and the prediction is, in fact, fulfilled. FLEDELIUS found a highly significant incidence of myopia amongst premature individuals even after puberty [6].

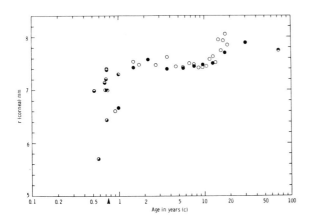

Fig. 4 The variation of the corneal radius of curvature [9]. Different symbols relate to different studies. Age since conception

Table 1 Mean ocular parameters for normal and premature emmetropic children of equal ages (data in mm).

	Refraction (D)	Axial length	A.C. depth	Lens thickness	Corneal curvature
Premature boys	+0.56	23.41	3.84	3.56	7.74
Full-term boys	+0.62	23.57	3.93	3.56	7.89
Premature girls	+0.52	22.77	3.81	3.65	7.61
Full-term girls	+0.59	23.41	3.89	3.55	7.81

If the assessment of the anatomical situation is correct then we are faced with a new problem. There is, as we noted, no suggestion that the neural development of the premature eye is retarded. On the contrary, one has to re-emphasize that the retina is able to function long before the normally required time. But the myopia of the premature bears the marks of arrested ocular development. Insofar as the ametropia is permanent, there appears to be no mechanism whereby the body can correct this.

It will be appreciated that if the prevalence of myopia amongst the prematurely born is between 15% and 20%, and, in the West, the incidence of prematurity is 6%-7% [7], then 1% of the total population is myopic probably because of being prematurely born. If this type of myopia should turn out to be preventible, national statistics for myopia will need reconsidering.

We face the question of whether there is a possible remedy available, e.g. in an environmental change. But before an attempt is made to answer this we ought to examine the third item on our list.

3. What Does Birth Do for Our Eyes?

There are two dramatic changes affecting the eyes at the moment of birth. One is a drop in the temperature of the anterior segment. As is well known, in adults the size of the core of the body over which the temperature is kept constant varies with the environmental temperature: it is greater in hot than in cold climates. In adults, the depth from the surface up to which considerable variations caused by the environmental temperature are detected is of the order of 1 cm [8]. In babies, in whom the surface/volume ratio is clearly much greater than in adults, temperature control is worse and even if the depth of only 1 cm were to hold (and it does not), it is effectively much greater than in the adult. Superficial tissues, which the eyes obviously belong to in this instance, consequently suffer thermal shock. But while the body is wrapped the face remains exposed. This is not the place to discuss thermoregulation in the infant: for the first years of life it is inefficient, and the eyes with their sparse blood supply face a problem much harder to surmount in the infant than in the adult.

A drop in temperature could be expected to lead to a local increase in the metabolic rate with a concomitant drop in the rate of growth owing to greater oxygen requirements. There is some evidence that this occurs both in the lens mass (Fig. 5) and in that of the eyeball as a whole (Fig. 6). The ex-

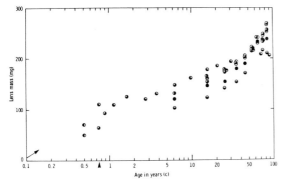

Fig. 5 The variation of lens mass with age [9]. Different symbols relate to different studies. Age since conception

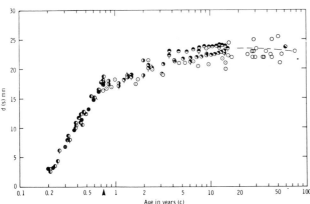

Fig. 6 The variation of the sagittal diameter of the eye-ball with age [9]. Different symbols relate to differential studies. Age since conception. Women: tailed symbols

ponential growth of the foetal eye is arrested at birth. Note that the period of resumed growth appears to coincide with the age during which efficient thermoregulation in being achieved, although the data available for its growth are much sparser. This seems to be true also for the crystalline lens. It is worth recalling that the lens appears to be growing continuously [cf. 9], and that a reduction in temperature (in the rabbit) may lead to mitotic rarefaction [cf. 10].

The other major event that befalls the eyes of the newborn is that they are exposed to light. There are several pointers to suggest that in one form or another, this may play a role. For example HOYT et al. [11] observed large negative errors of refraction in unilaterally closed eyes of neonates when the closures extended from 7 weeks to 1½ years. Even the shortest closures gave rise to axial elongation. This is not, however, a property of only young eyes since O'LEARY and MILLODOT [12] were able to correlate unilateral ptosis with ipsilateral myopia. There appears to exist also a relation between the degree of myopia and the sleep/wake ratio. Evidently, lid-closure makes it hard to distinguish between photic and thermal causes, both of which may, of course, play a role independently of each other.

Darkness appears to act in promoting myopia. Following prolonged periods in reduced illumination, both avian and mammalian eyes exhibit a marked in-

Table 2. Truth table relating environment and myopia

	Lower temp	Light reduction	Greater growth rate	More myopia
Prematurity	Yes	No	No	Yes
Full term	Yes	No	No	No
Ptosis	No	Yes	Yes	Yes
Experiment (Bercovitz et al., 1972)	No	Yes	Yes	Yes

crease in the size of the eye-ball [13]. Compensatory reductions in lens thickness and corneal curvature fail to stem the myopic tide that accompanies this malformation.

A truth table illustrates the difficulties facing one in trying to arrive at an answer. Table 2 shows a comparison of the environmental factors of light and temperature in promoting myopia in the experimental situation mentioned above and in prematurity respectively.

On the face of it, the myopia of prematurity [6] cannot be related to any of the experimental conditions. However, it has to be remembered that in the latter, we are dealing with fully developed eyes and that their defects are axial. But in prematurity, the eye is, if anything, too small and the myopia appears to result from corneal under-development and lenticular position (which is, of course, also an axial defect). The Table suggests, therefore, that the myopia of prematurity is sui generis. It also points to light (or contrast) reduction as a potent factor in the aetiology of some types of myopia.

As regards remedies, the suggestion that the myopia of prematurity may be characteristic of arrested development rather than akin, e.g. to that which accompanies ptosis may open up one or two possibilities. Although temperature is ruled out by Table 2 as a general factor, it may be important in the special case of prematurity: it is not normal to apply special heat to the premature face in incubation but the heat loss resulting from the face being at much less than 37° C during the time preceding full term may be significant. Directional radiators might provide a useful experimental tool, though pads might go some of the way. There is also the possibility that even with closed lids, premature eyes receive light sufficient to inhibit growth normally taking place in the darkness of the womb. Here again, padding might be of use.

4. Is There a Difference Between Men and Women?

A priori one would expect a difference between the eyes of the sexes just as there is one between body weight, bone composition, size of brain, etc. In this sense, it comes as no surprise if, at all ages, men's lenses are heavier than those of women [14]. This is a congenital difference and does not imply that there is a significant optical sex difference. It is important to distinguish such differences from others, perhaps due to hormones. Strictly speaking, these are, of course, also constitutional but they may manifest differently at different ages. In view of the several ectodermal components of the eye and of the fact that other ectodermally derived tissues, like hair

and skin, also reveal sex differences linked to growth, development and aging
[15], one has to consider the possibility that the eye, too, may show them.
In other words, as far as possible data from the two sexes should be kept
separate.

Allometric curves for the ocular components serve to emphasize the value
of such a procedure, and, at the same time, lead on to a noteworthy paradox.
The lens grows throughout life [9], as a result of which the depth of the
anterior chamber is reduced. One would expect, therefore, that, with women
having the smaller lenses, their anterior chambers would be correspondingly
deeper than those of men. But this is not observed. Just as the male lens
mass is about 4% greater than the female one, so is the depth of the anterior
chamber [9]. To a first approximation, these differences are constant throughout the life-span (Fig. 7). It would seem to follow that if both sexes have
similarly sized eye-balls, the refractions will differ if the refracting surfaces are similar. Before puberty, the male eye-balls are 2%-3% larger, a
difference that drops to an insignificant 1% thereafter. Thus women would
be expected to be slightly myopic in relation to men unless they compensate
for the intra-ocular difference with a larger corneal radius of curvature.
Adequate data distinguishing sexes from each other do not appear to be available. Moreover, SORSBY et al. [16] found "trivial" changes in corneal power
between the ages of 3 and 15 years: girls had somewhat smaller corneal radii
of curvature. E.g. at the age of 10, the difference in corneal refraction
amounted nearly to 1 D. The data both of BROWN [17] and of HIRSCH [18] are
consistent with the view that girls grow to be myopic in comparison with
equally old boys, and the results of FLEDELIUS [6] show this to be the case
on a statistically significant level: mature girls are almost 0.5 D less
hyper-metropic than boys (P<0.05).

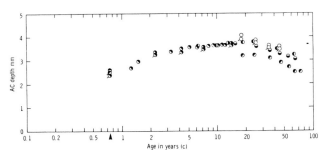

Fig. 7 The variation of the depth of the anterior chamber as a function of
age and sex [9]. Different symbols relate to different studies. Age since
conception. Women: tailed symbols

The optical differences between the two sexes appear to be established
even though their importance can be overestimated. There is, however, no
information on their micro-structural substrate, a matter of potentially
greater interest and significance. There are two pieces of evidence to suggest that there is a difference between male and female lenses other than
is attributable to their size and position. When human lenses of different
ages are compressed along their sagittal axis Hooke's Law is obeyed within
certain limits. Under compression, also, their birefringence changes. This
change is reversible for strains the magnitude of which varies inversely with
age [19]. Moreover, the strain to which women's lenses can safely be sub-
jected is significantly smaller than is true for those of men. A prediction
that follows from this, namely that women are at greater risk as regards se-

nile cataract is widely known clinically, and confirmed when suitable controls are applied [cf. 20]. If real, this difference is unlikely to be related to a global feature of the crystalline lens, such as its curvature. Rather is it going to be a matter of morphology (e.g. as regards the lenticular fibres) or perhaps even one of molecular anisotropy. Even if it were argued that the difference between the two types of lens only appears experimentally, e.g. because the lenses are handled, it is hard to sustain the notion that an experimenter will manipulate the two types of material so differently as to achieve the observed result.

The greater sensitivity of women's lenses to potential trauma surfaced also in a study concerned with image-forming characteristics. Measurements were made of the highest spatial frequency f(c) that lenses transmitted as a function of age, and days post mortem [21]. Within limits, neither of these parameters played a role (Fig. 8). However, the values of f(c) for men were consistently higher than those for women (P<0.02). This result fails to decide whether the difference between the two sexes becomes manifest only post mortem or whether it is present already during life.

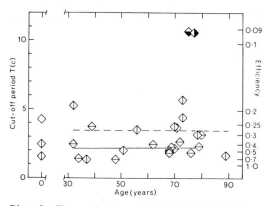

Fig. 8 The cut-off period [T(c) = 1/f(c)] of the crystalline lens as a function of age and sex [21]. The difference between the data fro the two sexes is statistically significant (P < 0.01). (⬦) Men; (⬦) Women; (⬦) brunescent cataract; (⬦) cloudy lens

A recent study [22] of the ocular and visual characteristics of large numbers of men and women within a wide age range offers little conclusive assistance, if any, because of the definition of the age groups. It shows nevertheless that, amongst those 65 and older, the percentage of women having vision better than 0.8 is smaller than is true of men (P<0.01). Below the age of 65 there is no significant difference. At present, one cannot rule out the possibility that this may be due to different rates of cell death in the two sexes [23]. This reservation notwithstanding, the results lend support to the message conveyed by this section, namely that potentially valuable information may be lost if data for men and women are averaged before they have been shown to belong to the same sample.

References

1   A. Marin-Amat: Bull. Soc. belge Ophthal. 113, 251 (1956)
2   V.M. Reading: The Contact Lens 4, 19 (1973)
3   O. Leser: Arch. d'ophthal. 42, 81 (1925)

4   C.J.W. Dunnewold: On the Campbell and Stiles-Crawford Effects and their Clinical Importance. Inst. for Perception RVO-TNO (1964)
5   D.M. Maurice: J. Physiol. 136, 236 (1957)
6   H. Fledelius: Acta Ophthal. 54, Suppl., 128 (1976)
7   R. Boldman and D.M. Reed: *Worldwide variations in low birth weight*. In the *Epidemiology of Prematurity*, ed. by D.M. Reed and F.J. Stanley (Urban & Schwarzenberg, Baltimore 1977)
8   B. Nielson: Acta Physiol. Scand. Suppl. 323 (1969)
9   R.A. Weale: *A Biography of the Eye: Development--Growth--Age* (H.K. Lewis & Co. Ltd., London 1982)
10 M.J. Voaden: Exp. Eye Res. 12, 328-336 (1971)
11 C.S. Hoyt, R.D. Stone, C. Fromer, and F.A. Billson: Am. J. Ophthalmol. 91, 197 (1981)
12 D.J. O'Leary and M. Millodot: Experentia 35, 1478 (1979)
13 A.B. Bercovitz, P.C. Harrison, and G.A. Leary: Vis. Res. 12, 1253 (1972)
14 J.J. Harding, K.C. Rixon, and F.H.C. Marriott: Exp. Eye Res. 25, 651 (1977)
15 Anon.: Brit. Med. J. 283, 1422 (1981)
16 A. Sorsby, B. Benjamin, M. Sheridan, J. Stone, and G.A. Leary: *Refraction and its Components During the Growth of the Eye from the Age of Three* (Spec. Rep. Ser. Med. Res. Counc., London, No. 301, H.M. Stationery Office 1961)
17 E.V.L. Brown: Arch. Ophthal. 19, 719 (1938)
18 M.J. Hirsch: Am. J. Optom. 29, 445 (1952)
19 R.A. Weale: Exp. Eye Res. 29, 449 (1979)
20 R.A. Weale: *Physical Changes due to Age and Cataract*, in *Mechanisms of Cataract Formation in the Human Lens*, ed. by G. Duncan (Academic Press, London 1981)
21 R.A. Weale: Exp. Eye. Res., in press (1983)
22 The Framingham Eye Study, Monograph. Survey of Ophthalmology, suppl. 24 (1980)
23 R.A. Weale: Trans. Ophthal. Soc. U.K. 95, 36 (1975)
24 E. Blechschmidt: *The Stages of Human Development before Birth* (S. Karger, Basel and New York 1961)
25 R.E. Scammon and E.L. Armstrong: J. Comp. Neurol. 38, 165 (165)

Wide-Angle Optical Model of the Eye

Oleg Pomerantzeff, Peter Dufault, and Robert Goldstein

Eye Research Institute of Retina Foundation, Boston, MA, 02114, USA

Several optical models of the eye have been proposed [1-6]. However, in all of them the distribution of refractive indices was taken either from biological or embryological data or from measurements on frozen sections of the lens. The optical performance of these models was a result of this design, and not its basis. Therefore, these models are not suitable for a study of the internal structure of the crystalline lens.

Our multilayer-structured model has been designed especially for calculating the distribution of refractive indices in the relaxed lens and for study of the redistribution of the indices in the process of accommodation.

Some parameters of the optics of the eye are measurable in vivo or in vitro. However, the internal structure of the crystalline lens cannot be measured in vivo and is distorted in vitro. To calculate this structure-- that is, the distribution of the refractive indices in the lens--we calculated an averaged curve of axial spherical aberrations of the emmetropic eye as a best fit to measurements on 100 eyes. This curve is then used for calculation of the unknown parameters of the lens by fitting the calculated curve of the model to the measured one. The design of this model and the assumptions used for it have been reported and discussed in 1971 and 1972 [7,8].

Two questions about this design must be resolved. 1) Is the solution obtained unique? 2) Is the model based on the curve of axial spherical aberrations valid for off-axis focusing?

To answer the first question we must review the design. The power of the cornea is completely defined by measurements (axial curvatures, thicknesses, and refractive index). The power of the crystalline lens has been selected from correlation studies by SORSBY et al. [9]. We assume that the refractive index varies continuously. This means that the number of layers should be infinitely large. However, for the study of accommodation, finite layered structure is more convenient. The number of layers was fixed at 200. With this number of layers the resulting discontinuities in the calculated aberrational curve are smaller than 5 μm (not resolved by the retina). For various experiments we have designed a subroutine that transforms the model to fit any number of layers. We also found that the number of layers does not affect the shape of the aberration curve, but only the size of the discontinuities. The general structure of the crystalline lens is shown in Fig. 1. If the front curvature RF, the back curvature RB, and the total thickness of the lens T are known, the ratio of sags

$$(TF + f)/(TF + B) = TF/TB = f/b = TFB$$

is easy to calculate. We provisionally assume that the height of the nucleus is very close to 1 mm. This assumption is based on the height of the

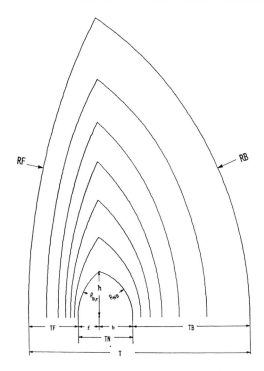 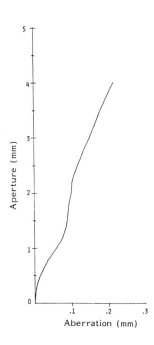

Fig. 1 Schematic of the internal structure of the model crystalline lens.
RF and RB are the measured radii of the anterior and posterior capsules,
respectively; T is the measured thickness of the entire lens; TF and TB are
the calculated thicknesses of the front and back parts of the lens, respectively, from the capsule to the nucleus; TN is the calculated thickness of
the nucleus; f and b are the calculated thicknesses of the front and back
sags of the nucleus, respectively; R_{NF} and R_{NB} are the calculated radii of
the anterior and posterior surfaces of the nucleus, respectively; h is the
equatorial radius (aperture) of the nucleus.

Fig. 2 Averaged observed curve of spherical aberrations. This curve is the
best fit for the measurements on 100 emmetropic eyes (50 volunteers).

inflection point in the aberrational curve (Fig. 2). The inflection is due
to the decrease of index in consecutive layers outside of the nucleus. The
value of this height (0.98 mm) will be adjusted exactly in the process of
fitting.

   We assume that the radii of the nucleus are in the same proportion RFB as
the front and back radii of the lens. Then from Fig. 1, it follows that

$$h^2 = f(2RF - f) = b(2RB - b) = 1$$

$$f/b = TFB$$

$$R_{NF}/R_{NB} = RF/RB = RFB \ .$$

   These four equations with the assumption h=1 mm define completely the
paraxial structure of the nucleus, that is, the values of R_{NF}, R_{NB}, f, and
b; hence the thickness of the nucleus TN = f + b.

13

Each layer is defined by its axial curvature, its axial thickness, its refractive index, and its aspheric coefficients. To reduce the number of unknowns, we assume that the physical parameters of the lens vary smoothly and monotonically from the front or back capsule to the nucleus. Therefore, we use a 3rd-order polynomial to define the parameters in either "half" of the lens.

$$P(j) = aj^3 + bj^2 + cj + d \quad , \qquad \begin{matrix} j=n \\ \\ j=1 \end{matrix}$$

where a, b, c, and d are the unknown coefficients, j is the order number of the layer, and n is the total number of layers. We assume also that the first and the second differences between the layers tend to zero at the nucleus. Therefore we write $P'(n) = P''(n) = 0$.

The distribution of axial curvatures is obtained separately in each half of the lens. Let RF and RB be the measured radii of axial curvature of the front and back capsule, respectively. We write four equations for each half.

$$RF = a_{RF} + b_{RF} + c_{RF} + d_{RF} \quad \text{for } j = 1$$
$$R_{NF} = a_{RF}n^3 + b_{RF}n^2 + c_{RF}n + d_{RF} \quad \text{for } j = n$$
$$R'_{NF} = 3a_{rf}n^2 + 2b_{RF}n + c_{RF} = 0$$
$$R''_{NF} = 3a_{RF}n + b_{RF} = 0 \quad .$$

These four equations for each half of the lens define unequivocally the four coefficients of the polynomial and therefore the distribution of axial curvatures in both halves of the lens.

Total thickness of the front part of the lens TF (not including the nucleus) is given by the equation:

$$TF = \sum_{j=1}^{j=n} [a_T j^3 + b_T j^2 = c_T j + d_T] \quad .$$

Axial thicknesses in the back part of the lens are obtained by dividing the values for the front part of the lens by TFB. To define the distribution of thicknesses of the lens, we can write the following equations:

$$T = (1 + \frac{1}{TFB}) \times \sum_{j=1}^{j=n} [a_T j^3 + b_T j^2 + c_T j + d_T] + TN \quad ,$$

where T is the total thickness of the lens, and T_{FC} is the thickness of the front capsule [10]. These four equations completely define the four unknowns a_T, b_T, c_T, and d_T.

The distribution of refractive indices among the layers is obtained in a similar way. This distribution is evidently the same on both sides of the nucleus. However, since we do not know the refractive index of the nucleus, we have only three equations. Let I_c be the index of the capsule [11].

$$I_c = a_I + b_I + c_I + d_I$$
$$3a_I n^2 + 2b_I n + c_I = 0$$
$$3a_I n + b_I = 0 \quad .$$

And we add $P_L = 20D$, where P_L is the power of the lens. At this point the paraxial portion of the system is unequivocally defined.

The off-axis shape of the cornea and lens, which is responsible for the aberrations of the system, is defined by fitting the aberrational curve. Our aspheric surface is defined by the equation:

$$Z = \frac{c\rho^2}{1 - \sqrt{1 + c^2\rho^2}} + as\rho^4 + bs\rho^6 + cs\rho^8 + cs\rho^{10}$$

where $\rho^2 = x^2 + y^2$ and $c = 1/R$.

We use four aspheric coefficients, as, bs, cs, and ds, each multiplied by the corresponding power of aperture. Each of these coefficients starts to affect the surface (and therefore the aberrations) at different increasing apertures. Thus, in the process of fitting of the curve each coefficient is introduced for fitting different progressively higher portions of the aberrational curve, as shown in Fig. 3. The extent of the effect of a given coefficient depends on the curvature of the surface. For the distribution of each of these coefficients among the layers of the lens, we use the same third-order polynomial and the same assumptions of smoothness and monotonicity. For the definition of four coefficients of the polynomial we have only two equations, $P'(j) = o$ and $P''(j) = 0$. Thus, remaining to be defined are two unknowns per coefficient and per half of the lens, or a total of 16 unknowns for the lens. We retain the four unknown coefficients for the front surface of the cornea, for a total of 20 unknowns. The effect of the back surface of the cornea amounts to about 4%-5% of the total power of the cornea.

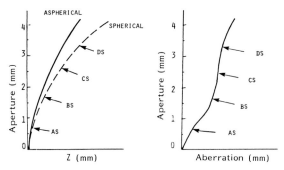

Fig. 3 Schematic representation of the effect of the aspheric coefficients on the spherical surface and the spherical aberration curve as a function of the aperture.

The variations of this effect in different eyes are negligible. We attribute only one aspheric coefficient to it and we calculate it only to match the measured thicknesses of the cornea at different heights.

If we could write an analytical equation giving the value of aberration at 20 different heights in the aperture in terms of the physical parameters of the lens--in this case, in terms of these 20 unknown aspheric coefficients--solving these equations would furnish the values of the unknown parameters. The calculated aberration curve would then match perfectly the measured curve in these 20 points. Taking another set of 20 apertures, we would match 20 other points of the curve. If our aberration curve is per-

15

fectly defined by 20 points and can be represented by the 19th-order polynomial, and if our system is also completely defined by the 20 parameters,
since these new points belong to the same aberration curve as defined by
the polynomial, the new set of parameters must be the same. Therefore,
this set of parameters is the unique solution. Actually, the system should
be defined by 1604 aspheric coefficients (800 for the 200 layers in each
half of the lens and 4 for the cornea). Reduction to 20 coefficients was
obtained by our assumptions. To check how well these assumptions define our
system, we select another set of 20 points and trace the aberration curve
for each set. These two curves are shown in Fig. 4. The difference between
the two curves is extremely small, a fact that we think can be considered
an argument in favor of our assumptions.

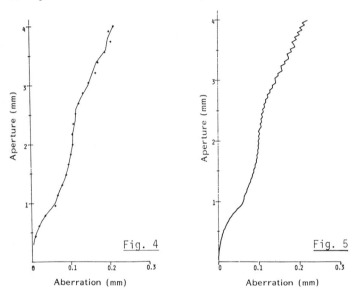

Fig. 4 Computer plot of the two spherical aberration curves (+'s and solid
line). Each curve represents calculated aberrations in a set of 20 points.
The aspheric coefficients were adjusted for one set of points and then the
same model was used to calculate aberrations in the second set of points.
The curves are practically identical.

Fig. 5 Computer plot of the spherical aberration curve of the emmetropic
model of the crystalline lens with 100 aperture points.

Since the analytical expression of aberrations is not possible, we define
numerical values of partial derivatives by ray tracing, and we use the method
of least squares to minimize the differences. The merit function of the fit
is the sum of squares of differences in all points. When this sum approaches
$.5 \times 10^{-3}$, changes in iterations become insignificant. We then trace 100
rays at 40 μm intervals. The result is a zigzag curve (Fig. 5). The maximum discontinuity is less than 3 μm, which cannot be resolved by the retina.
The goal of the fitting process is to fit the shape of the calculated curve
to that of the measured curve, whereas the zigzags, which reflect the discontinuities in the refractive index variation, are due to the finite dimension of layers. Increasing the number of layers, we reduce the discontinuities and at the limit would obtain a smooth curve. From this it follows

16

that the answer to the first question, "Is the solution obtained unique?" is YES.

The second question, regarding off-axis focusing, is answered by the same token. Indeed, we just showed that all the physical parameters of the system--axial separations, curvatures, indices, and aspheric coefficients--are unequivocally determined. The off-axis aberrations, as well as the axial aberrations, are all determined by those parameters and therefore cannot be different. The resulting model is shown in Fig. 6.

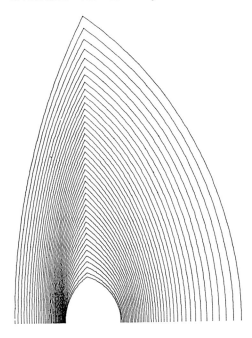

Fig. 6 Computer plot of the nonaccommodated model of the crystalline lens. Only 10% of the entire (398) set of the isoindicial layers are plotted

The sharp intersection of the front and back surfaces of the crystalline lens is an artifact of our computation. We calculate only the optically active portion of refractive surfaces up to 4.2 mm aperture. The plotting continues up to the intersection with the corresponding other half, using the curvature as adjusted at the last aperture. The equatorial portion of the lens is not used for focusing but only for attachment of zonules. Since our goal was to reproduce the performance and not the shape of the lens, we did not correct this portion of the lens. The lines inside the lens do not represent actual layers but isoindicial lines.

We compared the shape of the cornea we obtained by calculation with the average cornea resulting from very precise stereo-photogrammetric measurements by BONNET [12,13]. The result is shown in Fig. 7. The good match of these two curves is remarkable.

Accommodation

Mechanical changes in the crystalline lens in the process of accommodation, that is, change in axial curvatures of the back and front capsule and dis-

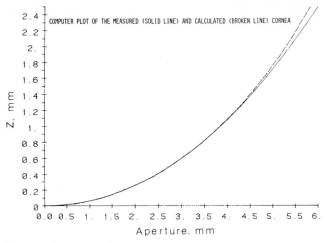

COMPUTER PLOT OF THE MEASURED (SOLID LINE) AND CALCULATED (BROKEN LINE) CORNEA

Z, mm

Aperture, mm

Fig. 7 Computer plot of the measured (solid line) and calculated (broken line) cornea. The final shape of the calculated cornea was derived in a process of fitting a calculated curve of spherical aberrations into the observed one.

placement of the lens, are taken from measurements by FINCHAM [14-16] and are introduced into our relaxed model. The new values of axial thicknesses and axial curvatures of the lens are calculated as for the paraxial portion of the relaxed state. The refractive indices of individual layers do not change. The aspheric coefficients are calculated as before but this calculation in the central portion of the lens up to the aperture of 1.5 mm (3-mm diameter) is done by forcing the image of the point at a distance of 250 mm from the pupil (least-blur circle of the focusing bundle of rays) to be located on the retina. For this calculation we use bundles of 100 rays originating from that object point and uniformly distributed on the pupil (3-mm diameters). This permits us to calculate the first two coefficients and their distribution (eight parameters). Outside of the central portion of the lens, the calculation is based on adjustment of the volumes. First we adjust the two missing parameters of the anterior and posterior capsule using the incompressibility of both halves of the lens. Then we calculate the two missing coefficients of the distribution on each side of the nucleus, by using the incompressibility of volumes in each isoindicial layer.

The result of this computation is shown in Fig. 8. Both the relaxed and the accommodated state are plotted by the computer, showing the change in lens shape and displacement. The resulting calculated curve of aberration is shown in Fig. 9. The axial distribution of refractive indices in relaxed and accommodated states is shown in Fig. 10, and the radial equatorial distribution is shown in Fig. 11. Variation of the index in any plane perpendicular to one of the axes (Y,Z) can be easily plotted.

This model in its relaxed state has been used for the design of various instruments; the results confirmed the validity of the model. Our goal, however, is to use this model as a base for designing individual models fitting any particular eye. This individual model must be automatically obtained by feeding into the program some measurements that can be easily obtained clinically. Individual models will have a variety of clinical and theoretical applications, such as design of intraocular lenses, exact measurements in the fundus, and design of low-vision aids.

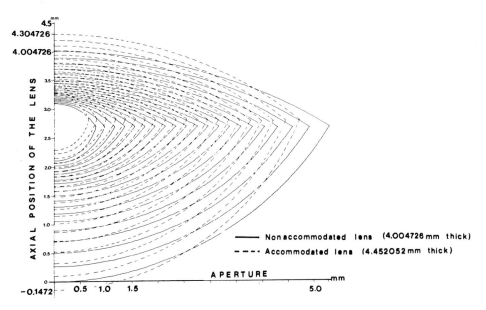

Fig. 8 Computer plot of the nonaccommodated (solid line) and +4D accommodated (broken line) crystalline lens models. Only 5% of the entire (398)
isoindicial layers are plotted. The initial position and the thickness of
the crystalline lens were changed when the lens went from the relaxed, nonaccommodated state into accommodation. The thickness changed from 4.004726
to 4.452052 mm, and the anterior surface was displaced forward 0.3 mm.

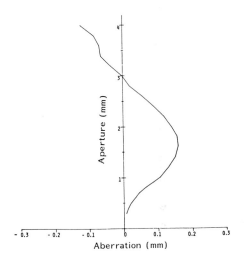

Fig. 9 Spherical aberration curve of the +4D accommodated model of the eye.

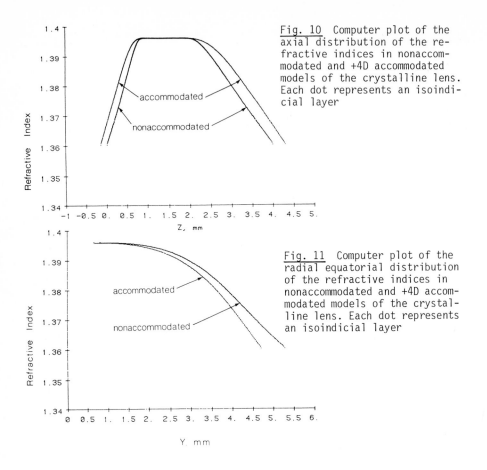

Fig. 10 Computer plot of the axial distribution of the refractive indices in nonaccommodated and +4D accommodated models of the crystalline lens. Each dot represents an isoindicial layer

Fig. 11 Computer plot of the radial equatorial distribution of the refractive indices in nonaccommodated and +4D accommodated models of the crystalline lens. Each dot represents an isoindicial layer

Acknowledgments

M. Pankratov for computations and drafting, and Dr. G.J. Wang for computations.

References

1 W. Lotmar: J. Opt. Soc. Am. 61, 1522 (1971)
2 T. Ono: J. Nara Med. Assoc. 20, 273 (1969)
3 S. Nakao and T. Ono: Jpn. J. Clin. Ophthalmol. 23, 41 (1969)
4 J.W. Blaker: J. Opt. Soc. Am. 70, 220 (1980)
5 N. Drasdo and C.W. Fowler: Br. J. Ophthalmol. 58, 709 (1974)
6 F.W. Fitzke: In: Modeling the Eye with Gradient Index Optics, ed. by A. Hughes (Cambridge University Press, Cambridge, in press)
7 O. Pomerantzeff, H. Fish, J. Govignon, and C.L. Schepens: Ann. Ophthalmol. 3, 815 (1971)
8 O. Pomerantzeff, H. Fish, J. Govignon, and C.L. Schepens: Optica Acta 19, 387 (1972)
9 A. Sorsby, B. Benjamin, J.B. Davey, M. Sheridan, and J.M. Tanner: Emmetropia and its aberrations. A study in the correlation of the optical components of the Eye. Medical Research Council Special Report Series, Her Majesty's Stationery Office, #293, London (1957)

10 R.F. Fisher: J. Physiol. <u>201</u>, 21 (1969)
11 F.P. Fisher, J.G. Vanmannen, J.S. Friedenwald, A. Sorsby, K. Steindorff, and W.W. Weisbach: *Tabule Biologicae*, Gravenhage: Uitgeverij. Dr. W. Junk, vol. 22 (1963)
12 R. Bonnet and P. Cochet: Bull. Soc. Ophthal. Fr. <u>73</u>, 688 (1960)
13 R. Bonnet: *La Topographie Cornéene*. Desroches, Paris (1964)
14 E.F. Fincham: Trans. Opt. Soc. <u>26</u>, 239 (1924)
15 E.F. Fincham: Br. J. Ophthalmol. 8, Monograph Supplement VIII (1937)
16 E.F. Fincham: Accommodation, in *Modern Trends in Ophthalmology* (F. Ridley and A. Sorsby, eds.), Butterworth and Co., London, pp. 268-275 (1940)

Measurement of Visual Axis Using a Laser Beam

Hiroshi Uozato, Hiroyuki Makino, Mototsugu Saishin, and Shuitsu Nakao

Department of Ophthalmology, Nara Medical University
Kashihara-shi, Nara-ken, 634, Japan

1. Introduction

A number of axes are recognized in the eye, such as optic axis, visual axis,
pupillary axis, fixation axis, and line of sight [1,2], as shown in Fig. 1.
It is well known that the visual axis is the most important for ophthalmic
prescription, measurement of refraction and corneal configuration and so on
[3] (Fig. 2). The visual axis is defined as the two lines, one from the
fixation point to the first nodal point in object space, and other from the
fovea to the second nodal point in image space [1]. However, objective
measurement of the visual axis is extremely difficult because it is imagin-
ary. In clinical viewpoint, the pupillary axis and line of sight are well
in use because of the simplicity of measurement. Strictly speaking, these
axes differ fairly from the visual axis, and do not pass through the fovea.

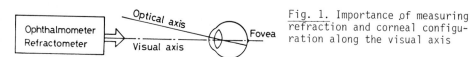

Fig. 1. Importance of measuring
refraction and corneal configuration along the visual axis

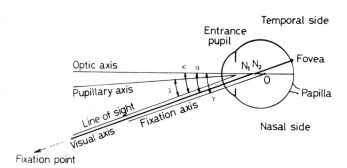

Fig. 2. Schematic representation of the axes and angles of the eye. N_1 and
N_2 are the first and second nodal points. O is the center of rotation. The
angles are exaggerated

   In this paper, we propose a new method for the precise measurement of the
visual axis of the human eye employing the directionality of a laser beam
[3]. We also suggest new techniques for adjusting the eye direction, which
can be used for refraction tests, photokeratometry, etc. Using the new procedure accurate examination of refraction and cornea curvature along the

visual axis of the patient's eyes may be easily realized. The principle and procedure of this method are discussed in detail, and its application particularly to objective refraction is also described.

2. Principle and Method

A schematic diagram of the optical system used is shown in Fig. 3. The collimated narrow laser beam emerges from a mirror with a pinhole incident upon the subject's eye, and falls on the fovea. This mirror was arranged to be perpendicular precisely to the laser beam axis. The subject aligns roughly by looking at his own eye's image formed by the mirror. Next, he fixates the cross hair located in front of the subject or its mirror image. The relative position of the visual axis to a laser beam spot in the object space can be monitored by a commercially available video camera or photographed by a camera which is focused on the pupil plane of the subject's eye. A laser beam reflected by a half-mirror is superimposed on the pupil image. Two optical paths, one from the pupil to the recording plane, the other from a laser source to the recording plane, are arranged to be coaxial.

When a subject fixates the cross hair target on a laser beam, and the cross hair and its mirror image are to be aligned, the visual axis in the object space passing through the entrance pupil plane can be obtained by the image of the pupil superimposing a laser beam spot, as shown in Fig. 4. Both horizontal and vertical displacement amounts relative to the center of the entrance pupil can be estimated from the enlarged image of the pupil.

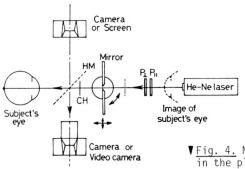

Fig. 3. Schematic diagram of optical arrangement used for measuring the visual axis. HM: Half mirror, CH: Cross hair, $P_{||}$, $P_{\perp}$: Two orthogonal polarizers

▼Fig. 4. Method of measuring the visual axis in the plane of the entrance pupil

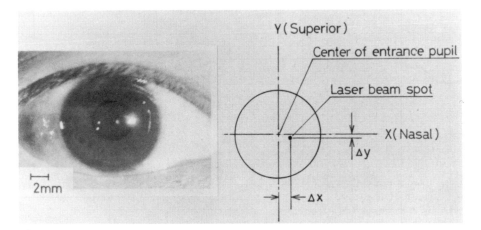

3. Experimental Results and Discussions

To demonstrate the feasibility of this method, simple experiments on a group of normal subjects were performed. A low power He-Ne laser (output power: 0.6 mW) was used as a light source. For eye safety from a laser beam, two orthogonal polarizers were used as an intensity control filter. Reduction ratio was about 10^{-3}. A plane mirror with a small hole (diameter: 0.2 mm) was located between the light source and the subject to be examined. This mirror, situated at 0.5 m in front of the subject's eye, was mounted on a movable stage (x-y translation and rotary) with micrometer heads, and arranged to be precisely perpendicular to the laser beam axis. The target of a cross hair, whose diameter was about 50 μm, was located at 40 cm in front of the subject's eye. Therefore, he could fixate the cross hair and a laser beam spot in the image of his own eye. Then the cross hair, its mirror image, and a laser beam spot were overlapped correctly on the subject's retina. When these three images were recognized to align by subjective judgment, the subject pushed the switch of the LED lamp. This signal was also recorded by the camera of the video monitor. A commercially available video system (Sony Betamax) and a monitor TV (14 inch) were utilized for recording and monitoring. Analysis was performed on the enlarged screen (20x) of the monitor TV or enlarged image of the photographed negative by using a profile projector (Nikon Model 6C) at 100x magnification.

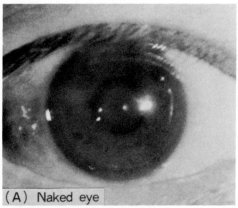

(A) Naked eye

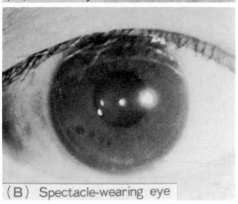

(B) Spectacle-wearing eye

Fig. 5a,b. Examples of the experimental results obtained in the states of naked eye (a) and spectacle-wearing eye (b), respectively

Figure 5 shows two examples; (A) was obtained with a naked eye and (B) obtained in a state of spectacle-wearing. This difference is mainly due to a prismatic effect of the spectacle lens. Relative positions of the visual axis to the center of the entrance pupil of the eye are shown in Fig. 6. Each of the data is a mean value of three measurements, and the solid bar represents standard deviation. The co-ordinates (0,0) correspond to the center of the entrance pupil of the eyes. The positive abscissa corresponds to the nasal direction, negative abscissa temporal direction, and positive ordinate superior, negative ordinate inferior directions, respectively. These results show that the visual axis passes through nasally and inferiorly to the center of the entrance pupil. This eccentricity of the visual axis is on an average 0.33 mm nasally, and 0.087 mm inferiorly, respectively.

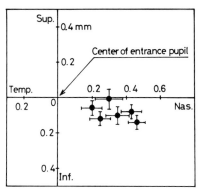

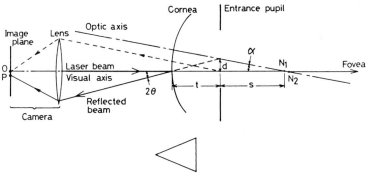

Fig. 6. Relative positions of the visual axis to the center of entrance pupil, obtained from six normal subjects

▼Fig. 7. Measuring method of the angle α by using a first Purkinje's image and a laser beam spot

As shown in Fig. 7, the angle α can be easily determined by using the first Purkinje's image and a reference point of a laser beam spot. Since the image plane is conjugate to the entrance pupil plane, the first Purkinje's image formed by a reflected laser beam at the corneal surface is obtained at slight displaced position P relative to the laser beam spot 0. In this case, however, the angle 2θ must be small enough to pass through into the recording lens aperture. In general, the distance PO in the image place is small, therefore the accuracy of its measurement is not sufficient. Fig. 8 shows another method to obtain the direction of the first Purkinje's image by using the direct detection of the reflected laser beam at the corneal surface. When the distance from the cornea to the detecting plane is ℓ, the reflected beam is displaced (=2·ℓ·θ) from the center of reference laser beam position. Therefore, the angle α can be computed by using the equation

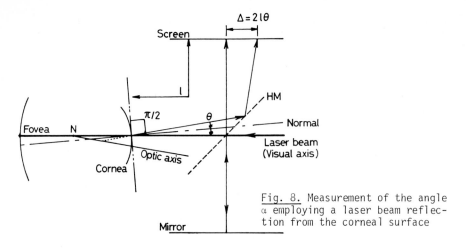

Fig. 8. Measurement of the angle α employing a laser beam reflection from the corneal surface

$\alpha = t \cdot \Delta / s \cdot \ell$, where t stands for the distance from the corneal apex to the entrance pupil, and s the distance from the entrance pupil to the first nodal point. The normal on the corneal apex on which the laser beam is incident can also be determined by this method.

If we use the values of Gullstrand's schematic eye, the entrance pupil and first nodal planes locate 3.05 mm and 7.08 mm apart from the corneal apex, respectively. Then the angle α can be estimated as 3.67° in horizontal and 0.97° in vertical meridians on an average, respectively. Table 1 shows the angle α in horizontal and vertical meridians obtained from six subjects. Each of the data was a mean value of three measurements.

Table 1. The angle α in horizontal and vertical meridians obtained from six subjects. Each data was a mean value of three measurements

Subject No.	Refraction	ANGLE α [deg.]	
		Horizontal	Vertical
1	S-0.50 ⊃ C-0.37 A170°	4.91	1.49
2	S-4.00 ⊃ C-0.75 A 30°	3.95	1.09
3	S-3.50 ⊃ C-0.50 A140°	3.40	0.09
4	S-4.50 ⊃ C-0.50 A160°	2.85	1.35
5	S-6.00 ⊃ C-0.37 A 20°	2.39	0.76
6	S-1.50 ⊃ C-1.00 A 5°	4.50	1.02
Mean and S.D.		3.67 / 0.96	0.97 / 0.48

As mentioned above, the visual axis is a very important problem in an objective refraction test and in measurement of the corneal configuration. The influence of the visual axis on the stability of objective refraction has been investigated [4]. Measurements of refractive data (sphere, cylinder and axis) were obtained along the visual axis and at a number of misalignment angles (2°, 4°, 6°) and measuring directions (temporal, nasal, superior and inferior) with the aid of a commercially available auto-refractor (Canon

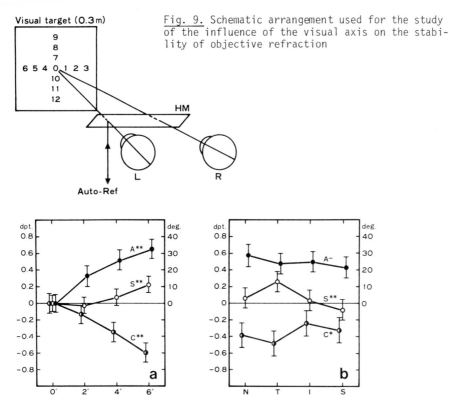

Fig. 9. Schematic arrangement used for the study of the influence of the visual axis on the stability of objective refraction

Fig. 10a,b. The effects of (a) the degree of misalignment angles and (b) misalignment directions. (**) Significant at 1% level, (*) significant at 5% level, (N) nasal, (T) temporal, (S) superior, (I) inferior

Auto-Ref R-1), as shown in Fig. 9.   Figure 10 shows the experimental results; (A) is the effect of the degree of misalignment angles and (B) the effect of misalignment directions. These results suggest that the coincidence between visual direction and measuring direction is very important for accurate objective refractioning, and that the variation of refractive data becomes large with the increase of misalignment angles. Therefore, the coincidence between the visual axis and measuring direction within 4° in arc is necessary for clinical examination.

For this purpose, an alignment device as shown in Fig. 11 is very useful for an accurate objective refraction test [4] or measurement of corneal configuration [5,6]. By using this technique, we can adjust the eye direction precisely on the visual axis. With the aid of a compact laser (e.g., He-Ne laser) or a visible semiconductor laser, refractometer or photoderatometer built-in this system can be easily realized.

Further applications of this method may be available to the alignment and fixation device, measuring the visual directions and the angles of the eye, and center of rotation (sighting center), etc. (Table 2).

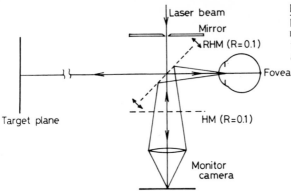

Fig. 11. Optical system applied to refraction test or measurement of corneal configuration along the visual axis

Target plane

Table 2. Some applications of the present method

1. ALIGNMENT AND FIXATION DEVICES
2. AXES OF THE EYE
3. ANGLES OF THE EYE
4. CENTER OF ROTATION (SIGHTING CENTER)
5. OPHTHALMOMOMETER AND KERATOMETER
   (PHOTOKERATOMETER, VIDEOKERATOMETER)
6. REFRACTOMETER (OBJECTIVE TYPE)

4. Conclusion

We propose a new method for precise measurement of the visual axis of the human eye in object space employing the directionality of a laser beam. Some experiments were performed to demonstrate the feasibility of this method. This method can provide a position of visual axis in object space passing through the entrance the entrance pupil plane. The accuracy of this method is the order of 2×10^{-2} mm at the entrance pupil plane, and $7 \times 10^{-2\circ}$ in arc for the measurement of the angle . We also suggest a technique for adjusting the eye, which can be applied to refraction test and photokeratometry and so on. This method also may be available to the alignment and fixation device, measuring the visual direction and the angle of the eye, center of rotation (sighting center), etc.

Acknowledgments

The authors wish to thank Mr. T. Ikuno for his technical assistance. This work was partially supported by a Grant-in-Aid for Scientific Research from the Ministry of Education (H. Uozato: No. 57771111).

References

1   H. Solomons: *Binocular Vision, a Programed Text*, William Heinemann Medical Books Ltd., London, p. 28-35 (1978)
2   Y. Le Grand and S.G. El Hage: *Physiological Optics*, Springer Series in Optical Sciences, Vol. 13, Springer, Berlin, Heidelberg, New York, pp. 71-74 (1980)

3   H. Ouzato, H. Makino, M. Saishin, and S. Nakao: J. Ophthalmol. Opt. Sci. Jpn. $\underline{3}$, 46 (1982)/Preprint of the 17th annual meeting of the Japan Society of Ophthalmological Optics (Oct. 4, 1981/Tsukuba) p. 11.
4   H. Uozato, H. Makino, M. Saishin, and S. Nakao: To appear in J. Ophthalmol. Opt. Soc. Jpn. $\underline{4}$ (1983)
5   H. Uozato, S. Kamiya, M. Saishin, and K. Nakai: J. Nara Med. Assoc. $\underline{32}$, 215 (1981)
6   S. Murakami, M. Saishin, H. Uozato, S. Kamiya, and Y. Mizutani: J. Jpn. C.L. Soc. $\underline{23}$, 215 (1981)

A Study of the Effects of Bleaching on the Width and Index of Refraction on Goldfish Cone Outer Segments

D.K. Hudson and J.H. Scandrett
Department of Physics, Washington University, St. Louis, MO, USA

J.M. Enoch
School of Optometry, University of California, Berkeley, CA 94720, USA

M.E. Bernstein
Department of Neurology, State University of New York, Stony Brook, NY, USA

Introduction

Earlier work with frog rod outer segments [1] indicated a slight swelling after bleach (2% to 4% increase in diameter). In this study, we examine the physical changes in the outer segment of goldfish cones that accompany bleaching.

Experimental Procedure

Goldfish, 2 to 3 inches long, were dark-adapted overnight. The goldfish were anesthetized and the anterior portion of the eye removed. Cones were teased from the retina into the immersion medium and prepared for examination. Time was recorded from the first cut of the eye; and every effort was made to limit the elapsed time between dissection and examination to minimize artifacts in the preparations. The immersion medium was goldfish aqueous humor.

Dissection was performed in near total darkness with the aid of an infrared image converter, and all measurements were made at $\lambda = 826$ nm in the infrared to prevent bleaching by the measuring beam. Black and white infrared photographs were taken of individual cones through a Zeiss Jamin-Lebedeff interference microscope adapted for use in the infrared. An initial sequence of photographs, with and without the interference pattern, was followed by an intense white bleach, followed by another series of photographs.

Typical times from the start of dissection to finish of pictures were between 5 and 10 minutes. Cones identified as long single (LS) according to the classification introduced by STELL and LIGHTFOOT [2] were selected and analyzed. These type cones are usually red or green receptors.

Analysis

Two types of measurements were taken:
1) The shape of the outer segment was determined from non interference images by measuring the length of the outer segment and the width at two points, each fixed distances from the tip of the cone.
2) The fractional fringe shift of fringes crossing the outer segment was measured. This shift determined the difference in optical path length through the outer segment compared with that of an equal depth of the immersion medium.

A computer controller microdensitometer was used for the analysis. The instrument consisted of an Apple II computer connected to an optical image scanner. A cathode ray tube served as a point source light, positionable by the computer on a 4096 by 4096 grid. Light passing through any point on a

35 mm film was compared to the light from the source alone. The logarithm
of the ratio was digitized on a 0 to 4095 numerical scale. Displays of interference and noninterference images of a cone outer segment are shown in
Fig. 1.

All edge features of the outer segment were located using a "maximum
lateral disturbance" signal developed by an averaging procedure. At 50
equally spaced intervals along the sampling line, a transverse density scan
was taken with 5 points on each side. A transverse density gradient was
estimated with weighted sums, and the total of the absolute values of all
50 scans formed a single value associated with the position of the sampling
line. In Fig. 1a, note that along the edge of the outer segment, displaced
fringes result in alternate positive and negative gradients--hence, the nec-
essity of taking the absolute value of the gradient signals.

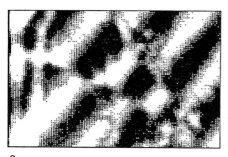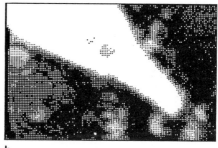

a   b

Fig. 1a,b. Images of cone outer segment (a) with interference pattern, (b)
(b) without interference pattern

The 50 values obtained by sweeping the sampling line across a cone outer
segment are shown in Fig. 2a. The two peaks are interpreted as defining the
cell width. Peak locations are computed by a least-squares fit. Figure 2b
shows the associated values of average density. A similar sweep down the
outer segment axis is shown in Fig. 3.

As in the previous report on frog rod outer segments, repeatability was
in the 1% to 2% range. Sigma/d for ten repeated width measurements on the
same image or of the same feature on different images were typically less
than 0.25%. A representative sigma is then less than 10 nm, which is very

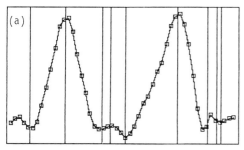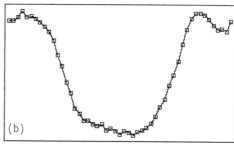

Fig. 2. (a) "Maximum lateral disturbance" scan across outer segment, (b) associated image density profile

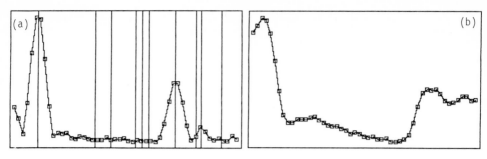

Fig. 3. (a) "Maximum lateral disturbance" scan down outer segment axis, (b) associated image density profile

Table 1 Goldfish LS cone outer segment measurements

Specimen	-4	-3	-2	-1	B L E A C H	1	2	3	4	% Change
C10-19-3										
Width 1		3.09	3.20	3.14		3.01	2.94	2.94		-4.7
Width 2		3.53	3.44	3.53		3.19	3.29	3.32		-6.7
Length	8.62								8.71	1.0
Time	5:41	5:44	5:46	5:48		6:04	6:06	6.08	6.11	
C10-23-1										
Width 1		2.23	2.24	2.20		2.23	2.20	2.26		0.3
Width 2		2.72	2.81	2.80		2.69	2.58	2.72		-4.1
Length	9.34								9.69	3.7
Time	6:02	6:06	6:08	6:10		6:27	6:29	6:31	6:46	
C10-24-2										
Width 1		3.57	3.46	3.55		2.75	2.77	2.78		-21.5
Width 2		3.73	3.81	3.77		2.82	2.84	2.84		-24.8
Length	7.31								7.12	-2.6
Time	5:41	5:45	5:47	5:49		6:06	6:08	6:11	6:14	
C9-18-1										
Width 1		2.77	2.44	2.58		2.51	2.50	2.63		-1.9
Width 2		2.99	2.88	3.01		2.81	2.75	3.00		-3.6
Length	7.93								8.19	3.3
Time	7:28	7:31	7:34	7:36		8:00	8:02	8:04	8:09	
C10-1-2										
Width 1		3.02	2.99	2.99		3.02	2.96	2.94		-0.9
Width 2		3.97	3.93	3.83		4.15	4.13	4.05		5.1
Length	10.71								10.78	0.7
Time	8:48	8:51	8:53	8:56		9:23	9:25	9:27	9:31	
C8-7-5										
Width 1		2.23	2.05	2.12		2.22	2.18	2.19		3.0
Width 2		2.30	2.31	2.31		2.26	2.25	2.25		-2.3
Length	11.51								10.97	-4.8
Time	3:37	3:40	3:42	3:44		4:04	4:06	4:08	4:11	

Mean % change after bleach
Width 1   -4.3±8.8
Width 2   -6.1±10.0
Length   0.2±3.0

Width 1 = width 1.92 microns from tip of outer segment
Width 2 = width 4.47 microns from tip of outer segment
Length = length of outer segment
Time   = elapsed time from first cut of eye in minutes:seconds

All dimensions in microns

Table 2 Goldfish LS cone outer segment index measurements

Specimen		BLEACH	
C10-19-3			
delta	.4047	A	.3968
delta/d	.1320	C	.1313
n_1	1.32952	H	1.32952
n_2	1.4615		1.4609
% change in delta/d = -0.5			
C10-23-2			
delta	.3767		.4087
delta/d	.1181		.1369
n_1	1.32975		1.32975
n_2	1.4478		1.4667
% change in delta/d = 15.9			
C10-24-2			
delta	.3565		.3480
delta/d	.1033		.1191
n_1	1.32955		1.32955
n_2	1.4329		1.4487
% change in delta/d = 15.3			
C9-18-1			
delta	.4496		.4421
delta/d	.1348		.1329
n_1	1.32945		1.32945
n_2	1.4620		1.4623
% change in delta/d = -1.4			
C10-1-2			
delta	.3979		.4029
delta/d	.1060		.1065
n_1	1.32966		1.32966
n_2	1.4356		1.4362
% change in delta/d = 0.5			
C-8-7-5			
fringes unreadable			

Mean % change in delta/d after bleach = 6.0±8.8

delta = optical path difference in microns
d = width of outer segment at fringe
n_1 = index of immersion medium
n_2 = derived index of outer segment

much less than the wavelength of illumination (826 nm). Hence measuring to less than a wavelength is not impossible, just difficult. The Rayleigh criterion sets a limit on image sharpness, but very small changes in the position of fuzzy edges can be determined by precise densitometric measurements and careful averaging procedures.

Fringe displacement was determined by scanning across displaced fringes in the center of the outer segment and scanning in a parallel direction across undisturbed fringes on each side. The fringe shift was obtained as the displacement of the center peak in a direction perpendicular to the line joining the two side peaks. The optical path difference (δ) is equal to the ratio of fringe shift (a) to fringe spacing (b) times the wavelength of light. The index of the outer segment (n_2) is given by

$$\delta = (n_2 - n_1) d,$$

$$n_2 = \frac{\delta}{d} + n_1 \; ,$$

where n_1 is the index of the immersion medium, and the thickness (d) is assumed to be equal to the width.

Results and Conclusions

Six available cone cells were found to be of the LS type. Measurements before and after bleach are summarized in Table 1. We observe that over a period of approximately 20 seconds the outer segment volume has decreased, on the average, approximately 10%, thus suggesting a bleach-related effect.

If the decrease in volume were attributed to loss of water, then the index of refraction would be expected to increase as is indicated in Table 2. Within errors, the changes of volume and outer segment refractive index are consistent with loss of water.

We will continue this study with goldfish rods and other types of goldfish cones and present a complete report.

Acknowledgments

This research has been supported in part by NEI Grant EY-03674 to J.M.E.

References

1 J.M. Enoch, J. Scandrett, and F.L. Tobey: Vis. Res. 13, 171 (1973)
2 W.K. Stell and D.O. Lightfoot: J. Comp. Neurol. 159, 473 (2975)

Part 2

Objective Refractometry

The Design of an Open View Autorefractometer

I. Matsumura, S. Maruyama, Y. Ishikawa, R. Hirano, K. Kobayashi, and
Y. Kohayakawa

Medical Equipment Design Department, Canon Inc., Nakahara-ku,
Kawasaki, 211, Japan

Introduction

Remarkable progress has been made in the design of automatic refractometers,
particularly because of advances in electronics and microcomputers. Various
types of automatic refractometers incorporating the advanced technologies
have been introduced and they have increased the demand for an instrument
that enables us to measure refraction with high reliability.

In order to achieve accurate refraction measurement, three goals have to
be achieved. (1) Instrument myopia can be eliminated. (2) The operation
can be performed easily. (3) A short measurement time, so that even patients
who tend to move can be examined accurately.

These three goals have been accomplished by using the following: (1) Measurement under natural viewing. (2) Alignment of the patient's eye on a TV
monitor. (3) Simultaneous measurement of three meridional directions (actual
measurement time: 0.2 sec).

Methods

The instrument consists of an alignment optical system and the measuring optical system. The principle of open view and alignment is shown in Fig. 1.
The patient's eye, looking at the fixation target through a dichroic mirror,
projects an image on the TV camera, i.e. an area of the pupil and cornea are
illuminated by the IR light sources. The reflected beams run toward the dichroic mirror, are reflected, and focused on the TV camera by the objective
lens.

Simultaneously, a ring-shaped alignment mark, illuminated by the light—
emitting diode, is projected on the TV camera by the projection lens. So,
the alignment can be accomplished by observing both the stationary mark and
the eye. The IR light sources are also used as targets. The optimum alignment between an eye and the measuring optics is made by adjusting the image
of light source to be clearly visible on a TV camera and then aligning the
edge of the pupil and an alignment mark so they appear concentric. Usually,
automatic refractometers are designed based on one of the following three
principles: Scheiner's principle, an image analyzing principle, or based on
retinoscopy. This instrument employs image-analyzing techniques.

As shown in Fig. 2, a beam from an indefinite source is refracted by the
toric lens at the point P ($r\cos\theta$, $r\sin\theta$) on the lens surface, and forms vertical line and horizontal line foci. A locus $P\alpha(X\alpha, Y\alpha)$ in the proper plane
at a distance of 1 D is introduced by the two formulae:

Fig. 1. Principle of open view and alignment

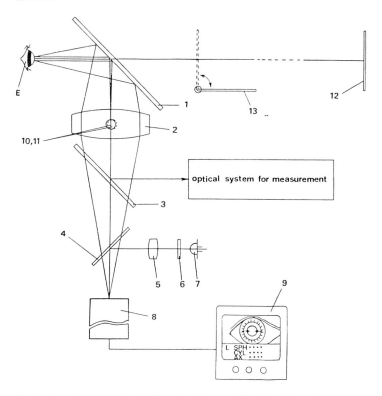

E: Patient's eye; 1: dichroic mirror; 2: objective lens;
3,4: dichroic mirror; 5: projection lens; 6: alignment
chart; 7: LED; 8: TV camera; 9: monitor TV; 10,11: light
source; 12: fixation target; 13: fixation target closed
by

$$\frac{r\cos(\theta - \alpha)}{\frac{1}{Dx}} = \frac{X\alpha}{\frac{1}{Dx} - \frac{1}{D}} \qquad (1)$$

$$\frac{r\sin(\theta - \alpha)}{\frac{1}{Dy}} = \frac{Y\alpha}{\frac{1}{Dy} - \frac{1}{D}} \qquad . \qquad (2)$$

From (1) and (2)

$$\frac{X\alpha^2}{Kx} + \frac{Y\alpha^2}{Ky} = 1 \quad ,$$

where

$$Kx = r^2(1 - \frac{Dx}{D})^2$$
$$Ky = r^2(1 - \frac{Dy}{D})^2 \quad .$$

Fig. 2: The image formation by the toric lens

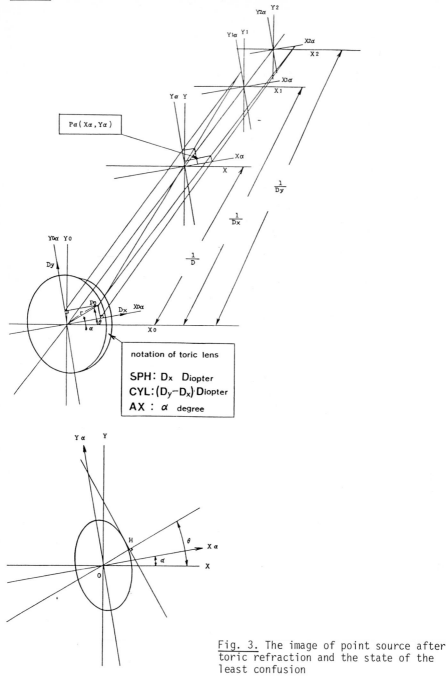

notation of toric lens

SPH: D_x Diopter
CYL: $(D_y - D_x)$ Diopter
AX : α degree

Fig. 3. The image of point source after toric refraction and the state of the least confusion

To obtain the spread of the point source in the direction $(\theta - \alpha)$, we consider the tangent line contacted with the ellipse shown in Fig. 3

$$Y\alpha = \frac{-1}{\tan(\theta - \alpha)} X\alpha + B \quad , \tag{3}$$

where

$$B = \pm\sqrt{Ky + \frac{Kx}{\tan^2(\theta - \alpha)}} \quad .$$

Since equation of the line with the direction $(\theta - \alpha)$ is given by:

$$Y\alpha = \{\tan(\theta - \alpha)\}X \tag{4}$$

from (3) and (4), the co-ordinates of the intersecting point become:

$$X\alpha = \frac{\pm\sqrt{Ky + \dfrac{Kx}{\tan^2(\theta - \alpha)}}}{1 + \dfrac{1}{\tan^2(\theta - \alpha)}} \tan(\theta - \alpha) \tag{5}$$

$$Y\alpha = \frac{\pm\sqrt{Ky + \dfrac{Kx}{\tan^2(\theta - \alpha)}}}{1 + \dfrac{1}{\tan^2(\theta - \alpha)}} \quad . \tag{6}$$

The spread of the point source accompanied by respective image distances after toric refraction are visualized in Fig. 4. The width of the confusion ellipse OH in the oblique direction (θ) is

$$OH = Ky \sin^2(\theta - \alpha) + Kx \cos^2(\theta - \alpha) \quad . \tag{7}$$

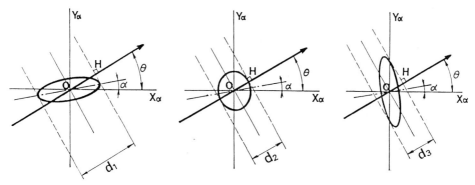

Fig. 4. Spread of the point source accompanied with the respective image distances after toric refraction [(d_1, d_2, d_3) blurred line image in the oblique meridian]

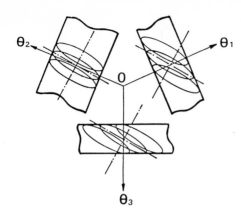

Fig. 5. Ellipsoidal sections through Sturm's Conoid related to each three meridians, and general formula of refraction

The ellipsoidal sections through Sturum's Conoid (θ_1, θ_2, θ_3) relative to each three different meridians along the line source are illustrated in Fig. 5. Now, the concept of the circle of least confusion is introduced as the general formula:

$$D = \frac{Dy^2 \sin^2(\theta - \alpha) + Dx^2 \cos^2(\theta - \alpha)}{Dy \sin^2(\theta - \alpha) + Dx^2 \cos(\theta - \alpha)} \quad . \tag{8}$$

Consequently spherical refraction, degree of astigmatism and axis of astigmatism of toric lens are found from the three formulae related to the detected power D_1, D_2, D_3.

$$D_1 = \frac{Dy^2 \sin^2(\theta_1 - \alpha) + Dx^2 \cos^2(\theta_1 - \alpha)}{Dy \sin^2(\theta_1 - \alpha) + Dx \cos^2(\theta_1 - \alpha)} \tag{9}$$

$$D_2 = \frac{Dy^2 \sin^2(\theta_2 - \alpha) + Dx^2 \cos^2(\theta_2 - \alpha)}{Dy \sin^2(\theta_2 - \alpha) + Dx \cos^2(\theta_2 - \alpha)} \tag{10}$$

$$D_3 = \frac{Dy^2 \sin^2(\theta_3 - \alpha) + Dx^2 \cos^2(\theta_3 - \alpha)}{Dy \sin^2(\theta_3 - \alpha) + Dx \cos^2(\theta_3 - \alpha)} \quad . \tag{11}$$

In Fig. 6, the beams emerge from the mask which consists of three meridians equally spaced from the optical axis and are projected to the fundus, passing through the focusing lens, relay lens, objective lens, and the central part of the cornea. Reflected beams from the fundus pass through the annular portion of the cornea, the objective and relay lens, and focusing lens. Then they are refocused on the detecting mask and finally received by the photodetector. The mask of the photodetector is conjugate with the detecting mask, and both are relative to the plane of the fundus. The focusing lenses are moved along the axis for power adjusting and each least blurred state is detected on a scale. Therefore, spherical visibility, degree of astigmatism and axis of astigmatism can be found by calculation from each position of the focusing lens at the time when the peaks are detected in the electrical outputs from the photodetectors. The results are presented on the TC monitor and can then be printed out.

In general, the objective lens, relay lens, and focusing lens are placed as shown in Fig. 6. The direct relation between the eye power (diopter) and the movement of the focusing lens (X) is given by:

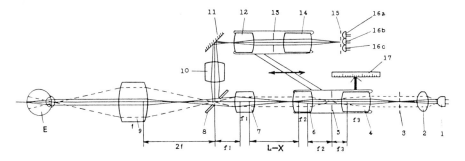

E: Patient's eye; 1: light source; 2: condenser len; 3: mask; 4,6,12,14: fo-
cusin lens; 5,13: field stop; 7,10: relay lens; 8: aperture mirror; 9: objective lens; 11: mirror; 15: detecting mask; 16: a~c detector; 17: scale;
f, f₁~f₃: focal length

Fig. 6. Schematic diagram of optical system for detecting the eye refraction

$$X = \frac{f_1^2 \cdot f_3^2 \cdot D}{1000(f_2^2 - f_3^2)} \; .$$ (12)

Results

The actual clinical evaluation of the instrument has been reported by many
doctors and investigators. The result of the Tokyo Medical and Dental University study is shown in Fig. 7. The excellent correlation between direc-
tive and subjective determinations show the instrument to be accurate and
reliable.

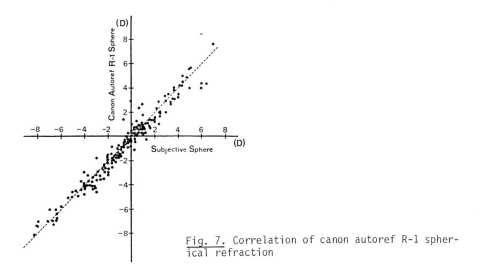

Fig. 7. Correlation of canon autoref R-1 spherical refraction

Summary

The design concepts of the automated infrared eye refractometer are described in this paper. This refractometer was modified so that it could be used in a natural viewing condition. With this design, the instrument myopia could be eliminated.

In order to shorten the measurement time, simultaneous measurement of three meridians is achieved, and the actual measurement time is 0.2 sec.

In addition, IR-TV is used to align and monitor the patient's eye, so that the operation can be performed by anyone easily. Some clinical evaluations of this instrument have been reported by various doctors and investigators with good agreement of the results by the subjective methods.

A Semi-Automatic Refractometer with a TV Monitor Allowing One Position Measures of Ametropia

Sukeyuki Oshima, Sachiko Hommura, Harumi Nose
Department of Ophthalmology, Institute of Clinical Medicine
University of Tsukuba, Ibaraki 305, Japan

Shinji Wada and Ikuo Kitao
Tokyo Optical Co., Tokyo, Japan

Introduction

Electronic refractometers have been used to measure ametropia objectively in
clinical practice for a decade. Focus detection is automated in many of them,
but there has been developed a unique instrument which allows television monitored measurement. The Topcon Refractometer RM 200B shown in Fig. 1 is
equipped with a TV monitor, on which the target patterns and the patient's
eye can be seen (Fig. 2). The target pattern consists of three pairs of bars
(Fig. 3), two of which are used for the detection of focus in the two principal meridians respectively, and the other is used for the estimation of
the axis of astigmatism. The usefulness of this particular design was investigated.

Methods

The principle of Topcon RM 200B is based on Scheiner's principle and developed from a coincidence-type refractometer. The optical system of Topcon
RM 200B is presented in Fig. 4. Figure 5 shows the target system, which allows the simultaneous measurement of refraction at two meridians perpendicular to each other. As distinct from refractometers of the telescopic type,
infrared rays are utilized for measurement and a television camera contained

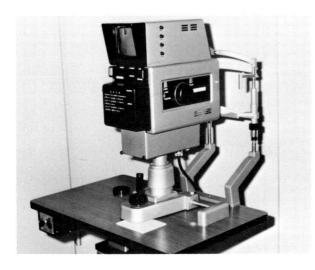

Fig. 1. Topcon refracto-
meter RM 200B

43

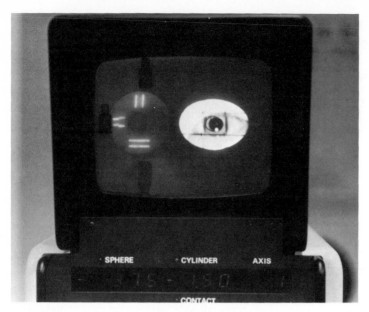

Fig. 2. The monitor screen on which the test targets and the patient's eye can be seen

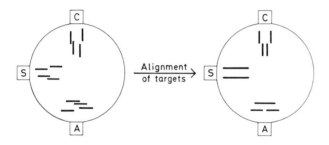

A : Target for axis detection

S : Target for focusing in 1st principal meridian

C : Target for focusing in 2nd principal meridian

Fig. 3. Illustration of test targets on the monitor screen of Topcon RM 200 B

in the instrument images the target which appears on the screen of a TV monitor, together with the subject's eye (as seen in Fig. 2). The relative energy of the wavelengths employed and the sensitivity of the TV camera are shown in Fig. 6. The dominant wavelength utilized for measurement was 854 nm.

The near infrared light gives no glare sensation nor after-image to the subject, and therefore the subjective refraction can be carried out immediately after the objective measurement. The procedure is similar to those using a full automatic refractometer, since it also provides the digital display on a panel and the printed record of the refraction data in terms of ophthalmic notation, i.e. sphere, cylinder, and axis which are obtained immediately.

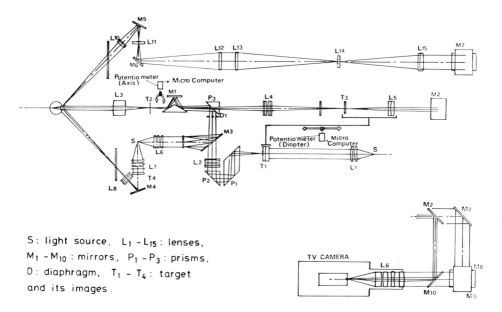

S: light source, L_1 - L_{15} : lenses,
M_1 - M_{10} : mirrors, P_1 - P_3 : prisms,
D: diaphragm, T_1 - T_4 : target
and its images.

Fig. 4. Optical system of Topcon RM 200B

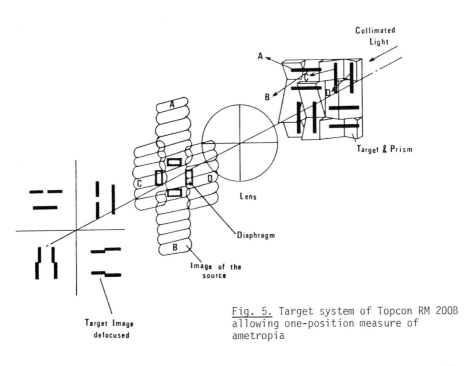

Fig. 5. Target system of Topcon RM 200B
allowing one-position measure of
ametropia

45

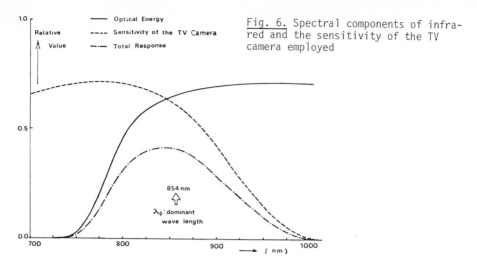

Fig. 6. Spectral components of infrared and the sensitivity of the TV camera employed

Subjects

Subjects in this study were 279 unselected patients who were examined in the University Hospital of Tsukuba. The ages of the patients ranged from 6 to 85 years old. Non-cycloplegic refraction of the eyes was measured with Topcon RM 200B and later, the subjective refractions were determined by a routine method utilizing a cross cylinder for the measurement of astigmatism. Four hundred and fifty-two eyes exhibiting a visual acuity of 1.0 or better with or without correction were chosen for the evaluation of the refraction data. Eyes with deteriorated vision were excluded since the subjective refraction might not be determined precisely.

Results

Now, two sets of refraction measurements were compared in terms of spherical equivalence, cylinder power and cylinder axis respectively. The correlation between the spherical equivalent data obtained by RM 200B and by subjective refracting is presented in Fig. 7. The two measurements are in general agreement over the entire range of ametropia. This agreement is evident from the high correlation coefficient of +0.976, as well as regression coefficients having a slope value of +0.933 and an intercept value of -0.079.

Comparison of cylinder power is shown in Fig. 8, where the astigmatism having the weakest principal meridian of from 45 to 134 degrees is arbitrarily denoted by plus value and the remainder by minus value. The correlation between the two sets of cylinder power data is fairly high, though slightly lower than that of spherical equivalence.

The determination of cylinder axis is shown in Fig. 9, where the discrepancies between the axis values are estimated by RM 200B and by subjective refraction for 300 eyes. Astigmatism of 0.25 D or more was detected with both methods. The discrepancy was computed by subtracting a RM 200B value in degrees from a subjective value, and it is clear from Fig. 9 that the larger the cylinder power the more precise the axis measurement. It is shown furthermore in Table 1 that the averaged discrepancy of the axis value amounted to -0.33° and the standard deviation was ±19.20° for 300 eyes in which astigmatism of at least 0.25 D was detected subjectively. The standard de-

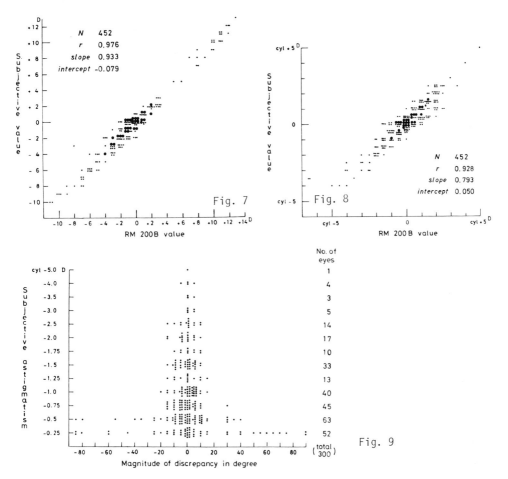

Fig. 7. Correlation between two measurements of equivalent sphere for 452 eyes; a large circle indicates 10 eyes and a small dot, one eye

Fig. 8. Correlation between two measurements of astigmatism in diopters for 452 eyes; a large circle indicates 10 eyes and a small dot, one eye

Fig. 9. Discrepancy between two measurements of the weakest meridian for 300 astigmatic eyes

viation decreased step by step as low amounts of astigmatism were excluded, though the mean values of discrepancies were not so different. Namely, the scattering of the axis discrepancy represented by standard deviation became remarkably narrow in proportion to the increase of cylinder power denoting astigmatism up to 0.75 D. A similar tendency was observed in a previously reported experiment [1], in which means and standard deviations were obtained from 20 successive measurements of model eyes having astigmatism from 0.12 D to 5.00 D, as shown in Fig. 10. It is suggested therefore that astigmatism less than 0.75 D may be sometimes falsely measured, since such measures are associated with inaccurate measurement of the axis.

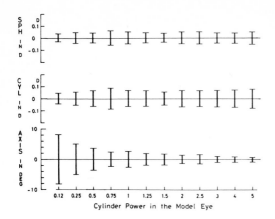

Fig. 10. Standard deviation in the measurement of model eyes with RM 200B

Cylinder Power in the Model Eye

Table 1 Discrepancy between cylinder axes estimated by Topcon RM 200B and by subjective refracting

Subjective astigmatism of	Discrepancy of axis value		No. of eyes
	Mean	S.D.	
cyl -1.00 D or more	-0.36 deg	± 6.59 deg	140
cyl -0.75 D or more	-0.62	± 6.99	185
cyl -0.50 D or more	-1.27	±12.42	248
cyl -0.25 D or more	-0.33	±19.20	300

Discussion

In order to obtain an accurate measurement of ametropia an objective refraction should be performed first, followed by a subjective refraction. If only the subjective refraction is done, the procedure is sometimes troublesome to the examiner and/or a patient. However, provided the refraction is preceded by the objective measurement, the subjective refraction will be done easily and efficiently because the trial lenses to be employed for the subjective refraction can be kept to a limited number, and the refraction determined precisely and in a short time. One reason why the subjective refraction may be a lengthy procedure is that the determination of astigmatism comprises two parameters, i.e. values of cylinder and axis which are independent of each other. Inaccuracy in the objective estimation of astigmatism cannot therefore simplify the subjective refracting procedure, and the usefulness of the objective refractometry will be lessened.

The accuracy of refractometry may be impaired by the fluctuation of accommodation, and the shifting of the eye in reference to the measuring beam, as happens during the shift of instrument position when measuring the different meridians using a refractometer of the manual type. Such a refractometer (usually of the telescopic type) may cause inaccuracy if the accommodative state of the patient's eye changes during the successive measurements. Therefore, the difference between the dioptric values at the two meridians may not represent a true amount of astigmatism.

Measurements in the different meridians are not necessarily made simultaneously in the case of an automated refractometer, though the procedure to obtain the measurements is somewhat different. The estimation of astigmatism with an automated refractometer is accomplished by computing dioptric values in at least three meridians which are inherent to the instrument and independent of the principal meridians in the examined eye. The computation is theoretically based on the sine squared relation indicating the dioptric value at each meridian of an astigmatic eye, as shown in Fig. 11. An example of the computation formulae reported by MALACARA [2] and WADA [3] is as follows: the axis value A (degrees), the cylinder power B (diopters) and the spherical value C (diopters) are given from the following three equations.

$$A = \frac{1}{2} \cot^{-1} [\frac{1.155 \, (P_{60} - P_0)}{P_{120} - P_{60}} + 0.557]$$

$$C = 1.155 \, (P_{120} - P_{60}) \, / \, \sin 2A$$

$$S = P_0 - (1 - \cos 2A) \, / \, .$$

Here, the refractive values at 0°, 60° and 120° meridian were denoted by P_0, P_{60} and P_{120} respectively.

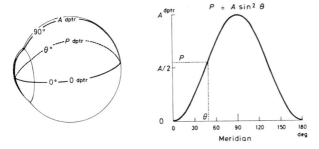

Fig. 11. Dioptric values of refraction at each meridian of an astigmatic eye (an example of cyl A D axis 180°). The dioptric values at 0° and 90° meridian are 0 D and A D respectively. Let the value at θ° meridian be denoted by P D, P is given by the expression $P = A \sin^2 \theta$

A specific example of the calculation is shown in the case of the astigmatism denoted by cyl -1.00 D axis 90°. When the astigmatic eye is measured with an electronic refractometer having an automated focusing device, it is expected from the sine-square relation that the dioptric value at the 0°, 60° and 120° meridians should be -1.00 D, -0.25 D, and -0.25 D respectively, as seen in the top line of Fig. 12. Now, strictly speaking the accommodation of the examined eye is not always constant during the measurement; minute fluctuations of accommodation may occur, voluntarily or involuntarily. Since measurements in each meridian cannot be made simultaneously, but in the case of refracting of a fully automated instrument are always made successively, it can be supposed that the eye accommodates by 0.1 D at the moment of measuring at the 120° meridian. Then the dioptric value at that meridian will be estimated as -0.35 D, and the calculated refraction value, derived from the above-mentioned equations, will be denoted by sph -0.06 D, cyl -0.94 D axis 86°, as seen in the second line of Fig. 12. If accommodation of 0.25 D or 0.5 D is elicited at the moment of measuring at the 120°

Measurements at 3 meridians   **Computed values**

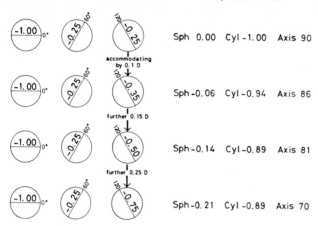

Sph 0.00 Cyl -1.00 Axis 90

Sph -0.06 Cyl -0.94 Axis 86

Sph -0.14 Cyl -0.89 Axis 81

Sph -0.21 Cyl -0.89 Axis 70

<u>Fig. 12.</u> An example of the computation of sph, cyl, and axis value from the measurements of dioptric refraction values at three meridians

meridian, the dioptric value at that meridian will be estimated to be -0.5 D or -0.75 D, and the calculated refraction will be sph 0.14 D with a cyl -0.89 D at axis 81° or sph -0.21 D, cyl -0.89 D axis 70° respectively, as exhibited in the third or bottom line of Fig. 12. Therefore, in a fully automatic instrument, even if the accommodation which occurs during measurement is a slight amount, the induced error of the measurement affects not only the dioptric values denoting the sphere and cylinder correction, but also the axis value.

The Topcon RM 200B employed in this study, however, allows the estimation of dioptric values in two principal meridians at the same time, the two meridians being watched on a TV monitor by an examiner. The mechanism is effective in accurate measures of astigmatism, because any fluctuation of accommodation during the measurement misaligns the bar target, a displacement easily differentiated from the configuration of the bars caused by astigmatic error. It was also useful (for precise measurement) to have the patient's eye projected on a monitor screen without interruption during the measurement. In this way, the eye can be watched together with the target patterns for measurement; shifting of a patient's eye can be checked and reliable data obtained. Although the Topcon RM 200B requires manual procedure it was far easier to operate the instrument than the ordinary manual telescopic refractometer. In the Topcon, the pictures of target patterns and the patient's eye are sufficiently bright to operate the instrument under normal room illumination (about 500 lux); all the images can be observed binocularly, unlike the monocular observation used in the telescopic system.

Although it was shown in this study that low astigmatism (less than 0.75 D) is sometimes falsely measured even with the Topcon RM 200B, such erroneous measurement may happen more frequently in the case of employing other types of objective refraction. It is, in any case, worthwhile to use an objective method prior to the subjective routine, in order to diminish the time spent with a patient, and also to improve the final accuracy in determining the refraction of the eye.

The Topcon RM 200B does not possess a fully automatic system but does incorporate the characteristic property allowing one-position refractometry. The latter provides a more reliable measurement of astigmatism than an automated focusing device, and seems to be useful in the refraction practice in spite of the development of the fully automated refractometer.

Conclusions

Experiences with a Topcon Refractometer RM 200B for objective refracting were reported. The instrument is equipped with a TV monitor on which a patient's eye and target patterns for measuring ametropia can be seen at a glance. The target patterns consist of three pairs of lines to be aligned; they are used for focusing in the first and second principal meridians and for detecting the axis of astigmatism. The measurement of ametropia can be accomplished without shifting the instrument position, that is, one-position refractometry can be carried out. The measurement data in standard ophthalmic notation are displayed on a panel and printed out immediately after the measuring procedure.

Four hundred and fifty-two eyes were refracted subjectively and also measured with the Topcon RM 200B. The refraction data obtained with both methods were compared in terms of equivalent sphere, cylinder power and axis, and the agreement of the two sets of data were good. It was far easier to operate the Topcon RM 200B than the ordinary manual telescopic refractometer. It was felt that the semi-automated refractometer, allowing one-position measure of ametropia, provides a more reliable measurement of astigmatism than does a fully automated instrument.

References

1 S. Oshima, S. Hommura, Y. Usuki, K. Kono, S. Wada, and I. Kitao: Journal of Ophthalmological Optics of Japan 2, 2 (1981)
2 D. Malacara: Am. J. Optom. & Phys. Optics 51, 15 (1974)
3 S. Wada and I. Kato: Rinsho Ganka 30, 1441 (1976)

A New Microprocessor-Assisted Objective Refractor

F. Corno, J. Corno, A. Roussel, and J. Simon

Institut d'Optique, Centre Universitaire, Orsay-Cedex, France

1. Introduction

Automated objective optometers (or refractors) enabling ophthalmic practitioners to measure ocular refraction automatically appeared within the last
decade [1-4]. Bausch and Lomb's ophthalmetron seems to have been the first
of these instruments, soon followed by Acuity System's Autorefractor and Coherent Radiation's Dioptron [5,6,7]. With this type of instrument, the operator looking at the patient's eye sets the eye-instrument distance and places
the pupil in a central position. Then ocular refraction is automatically
measured by the instrument which gives the prescription relative to the patient's eye as a print-out on a paper roll. An exhaustive review of automatic refraction instruments was published recently by GUYTON [8] and SASIENI
[9]. Automated refractors are convenient for they make it possible to find
which corrective lenses should be prescribed; but their purchase and main-
tenance costs are high. Moreover they deprive specialists of the opportunity
to observe the ophthalmoscopic image that would enable them to spot possible
ocular diseases. That was why we decided to conceive a simpler refractor
assisted by a microprocessor which avoids such drawbacks.

In this paper, a new instrument, the IOTA refractor, is described and its
performances are evaluated. The reliability of this refractor is assessed
by calculating the variance (or standard deviation) of a set of refraction
measurements repeated on the same eye (with the refractor). The validity of
refraction measurements is evaluated by comparing I.O.T.A. optometer refractions with those obtained by conventional subjective methods on the same
(large) group of subjects.

2. Principle and description of the instrument

This is the second version of a prototype working as a conventional refractor with a small angular test target. The specialist projects the test on
the patient's retina and observes the image. He can adjust the focus, measure the subject's ametropia very quickly and also study the quality of the
ophthalmoscopic image. An automated data processing device is associated
with the instrument and relieves the operator of tedious tasks which had to
be performed with conventional refractors: reading and interpolation of dioptric and orientation scales, identification of the examined eye, calculations, writing out prescriptions.

The instrument has the following characteristics:

Refraction measurement through analysis of the image quality of a punctum test in white light.

Full pupil measurement: the instrumental pupil fully covers the patient's pupil so as to make eccentricity tolerable without impairing measurements.

Simultaneous presentation of a measurement test and a relaxation test.

Use of a microprocessor which controls the various phases of measurement (centering, focusing...) takes the focusing values into account, computes refraction and displays it on LED, and shows the results printed on a paper roll.

Figure 1 presents a simplified diagram of the refractor in the case of a test focusing on the retina (hypermetropic eye). The test object T_0, luminance 4.10^{+6} cd.m^{-2}, is projected by the objective O_1 in T. The lens of the subject's eye forms an image T'_0 on the retina. The image of T'_0 reflected by the eye is again formed in T. Objective O_2 gives an image T''_0 of T observed by the practitioner through an eye piece (not shown in the figure). The instrument is devised so that the points T_0 and T''_0 are shifted synchronously.

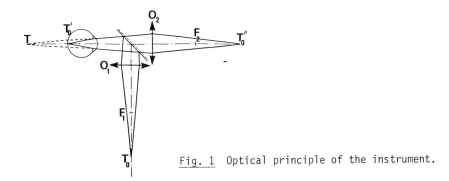

Fig. 1 Optical principle of the instrument.

A peripheral field (showing a landscape), luminance 900 cd.m^{-2}, is presented to the subject's eye so as to inhibit accommodation. To do so, it is necessary to start from a sufficiently hypermetropic position in order to prevent the subject from accommodating on the small target, the landscape becoming sharper and sharper as T'_0 gradually comes closer to the retina (fogging method).

Refractor automatization includes acquisition and treatment of data necessary to refraction computation as well as control of different measurement steps (resetting adjustment, focusing). These data are treated by a microprocessor whose power is quite sufficient to program and provide the following processes:

Resetting of the instrument, real-time computation and display of axis value and dioptric power of one or two focusings necessary to determine ametropia.

Adjusting, measuring of light sources and motors.

Printing results on a paper roll.

2.1 Characteristics of the prototype (Fig. 2)

Dioptric range from -20 to +15 D.

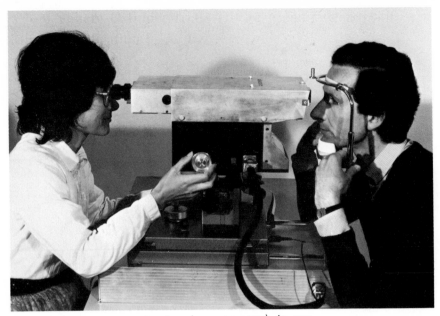

Fig. 2 The I.O.T.A. refractor (or optometer) in use.

Accuracy measurements: sphere ±0.25 D; for astigmatism above 0.5 D cylinder ±0.25 D, axis ±5°.

Global results given after measuring and examining second eye.

Automatically rounded dioptric values and either plus or minus cylinder transposition.

Reducing of measurement time to less than 60 seconds per patient.

3. Results

3.1 Accuracy of the Results

Refractor-measuring accuracy was assessed by calculating the standard deviation (S.D.) of a set of 50 focusing operations repeated on the same eye. Thus, reliability could be evaluated. Three eyes were examined without cyclopegia: the first was myopic and mildly astigmatic, the second myopic and heavily astigmatic, and the third practically emmetropic.

4. Comparison Between Subjective Refraction and Objective Refractor

Quantitative evaluation of this new instrument including a comparison with a clinical refraction shows the validity of measurements. In this comparison, half of the subjects examined were private patients (an ophthalmologist's private practice) and the other half, unselected people working in industry and at university. The monocular subjective refraction was possible on 164 eyes of 86 subjects, the objective examination with the I.O.T.A. refractor was performed on 158 eyes and the comparison could actually be made for 155 eyes.

4.1 Distribution of Ametropia

Results relate to values of subjective refraction which consists in setting
a spectacle lens at a distance of 12 mm from the cornea. They refer to the
155 eyes, for which the comparison between subjective examination and objective refraction was possible. The microprocessor of I.O.T.A. objective refractor permits the display of data in standard ophthalmic notation taking
the spectacle lens-cornea distance into account.

The measurement scatter given by subjective examination is about ±0.12 D
for subjects whose spherical ametropia is below 4 D or whose astigmatism is
below 2.5 D. For serious defects or patients with problems (amblyopia, etc.)
the scatter is higher.

Figure 3 shows histograms giving the frequency distribution of ametropia
among the subjects in this study. Ametropia for the 155 eyes included in
this analysis, stated in spherical equivalent form, ranges from -8 D myopia
to +15 D hyperopia. Figure 3a shows that in most cases, ametropia results
in myopia rather than hypermetropia. Therefore, the distribution of ametropia in this sample is slightly different from the classical distribution
showing a predominance of slight hypermetropia. The astigmatism of these
subjects (represented by the cylinder power component) ranges from 0 to 6 D
(Fig. 3b); the majority of these subjects has an astigmatism below 3 D. In
this paper, transposition to plus was chosen for the values of cylinder powers.

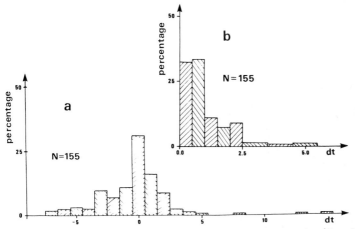

Fig. 3 Frequency distribution histogram of ametropia (N = 155 eyes)
a - ordinate: Frequency (percentage); abscissa: spherical equivalent refrac-
    tion, diopters (dt).
b - ordinate: Frequency (percentage); abscissa: cylinder power component,
    diopters (dt).

4.2 Comparison with Monocular Subjective Refraction

It is possible to make a comparison between the I.O.T.A. refractor and subjective refractions either in a global way, by analyzing the values of spherical equivalent refraction which partly takes astigmatism into account, or by
separately analyzing the three refraction components, namely spherical power,
power and axis of cylinder.

Results given here make a comparison between refractor and subjective refraction data possible. For this purpose it is necessary to consider:

1. The correlation between the subjective refraction (S) and the value given by the I.O.T.A. refractor (O) concerning the 155 pairs of data.

2. Cumulative frequency distribution of differences [S-O] between each pair of subjective-objective data for the 155 eyes.

5. Equivalent Sphere

The correlation between the I.O.T.A. refractor and subjective refraction spherical equivalent data is represented in Fig. 4 which shows that the two sets of measurements are in agreement. Figure 5 gives cumulative frequency distribution of differences [S-O] in spherical equivalent refraction between I.O.T.A. refractor and subjective refractions. Therefore this graph shows that 80% of the I.O.T.A. refractor measurements differ from those of subjective refraction by 0.50 D or less and that 90% differ by 0.75 D or less.

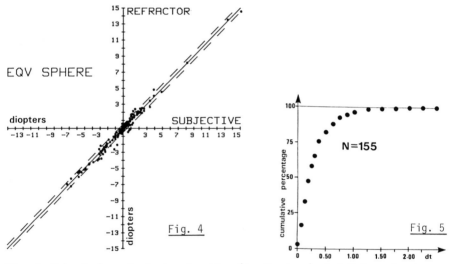

Fig. 4 Spherical equivalent refraction (in diopters): I.O.T.A. refractor (ordinate); subjective refractions (abscissa); N = 155 eyes. Coefficient of correction r_0 = 0.99. The regression line has a slope p_0 = 1.01 and intersects the abscissa at a value y_0 = -0.07 dt. The sample standard error or estimate: s' = 0.42 dt.

Fig. 5 Cumulative frequency distribution of differences in spherical equivalent refraction between I.O.T.A. refractor and subjective refractions (N = 155 eyes). Ordinate: Cumulative percentage; abscissa (diopters): difference [S-O].

6. Comparison of Refraction Components

The correlation between subjective examination and objective refractor is represented in Figs. 6 to 9.

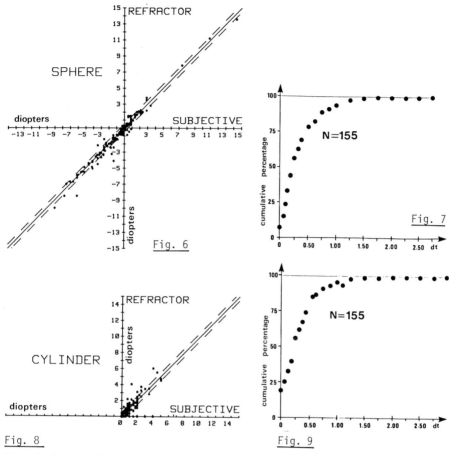

Fig. 6 Spherical power components (in diopters) of I.O.T.A. refractor (ordinate) and subjective refraction (abscissa); N = 155 eyes. Coefficient of
correlation r_0 = 0.98. Regression line has a shape p_0 = 1.02 and intersects
the abscissa o at y_0 = -0.07 dt. The standard error s' = 0.52 dt.

Fig. 7 Cumulative frequency distribution of differences in spherical power
components between I.O.T.A. refractor and subjective refractions (N = 155
eyes). Ordinate: cumulative percentage; abscissa (diopters): difference
[S-0].

Fig. 8 Cylinder power components (in diopters) of I.O.T.A. refractor (ordinate) and subjective refractions (abscissa). N = 155 eyes. Coefficient of
correlation: r_0 = 0.90. Regression line has a slope p_0 = 1.03 and intersects the abscissa at y_0 = 0.00 dt. The standard error of estimate: s' =
0.50 dt.

Fig. 9 Cumulative frequency distribution of differences in cylinder power
components between I.O.T.A. refractor and subjective refractions (N = 155
eyes). Ordinate: cumulative percentage; abscissa (diopters): difference
[S-0].

6.1 Sphere

Figure 6 shows regression line (full line) and two lines set at ±0.5 dt from this line (dashed line).

Figure 7 represents cumulative frequency distribution of differences [S-O] in spherical power component (diopters) between I.O.T.A. refractor and sub-jective refraction. Once again, 80% of comparisons differ by 0.50 D or less.

6.2 Cylinder

Figure 8 shows regression line and two parallel lines set at ±0.5 D. Cumu-lative frequency distribution of differences [S-O] in cylinder power compo-nent (diopters) shows that 80% of comparisons [S-O] differ by 0.50 D or less and 92% by 0.75 D or less (Fig. 9).

6.3 Axis Cylinder

For the subjective-objective comparisons of cylinder axis, all anastigmatic subjects (cylinder component = 0) and subjects whose astigmatism was lower than 0.5 D have been eliminated. Therefore comparisons have actually been made with 83 eyes (Figs. 10 and 11) whose astigmatism is for most of them below 3 D. Once again, results are in agreement. Fig. 10 shows correla-tions between the two sets of measurements (subjective-objective) for the 83 pairs of data. In this figure, line regression and two parallels placed at ±10 degrees are represented. Cumulative frequency distribution of dif-

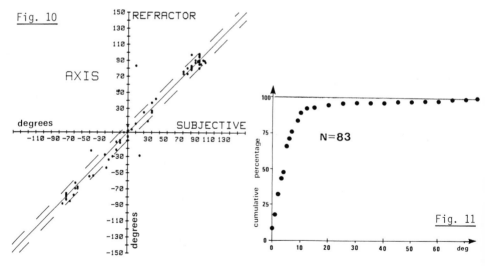

Fig. 10

Fig. 11

Fig. 10 Cylinder axis components (in degrees) of I.O.T.A. refractor (ordi-nate) and subjective refraction (abscissa) (N = 83 eyes). (Astigmatism > 0.5 diopter.) Coefficient of correlation r_0 = 0.98. Regression line has a slope p_0 = 1.00 and intersects the abscissa at value y_0 = 0.3 degrees. The sample standard error of estimate: s' = 13 degrees.

Fig. 11 Cumulative frequency distribution of differences in cylinder axis components between I.O.T.A. refractor and subjective refractions (N = 83 eyes). (Astigmatism ⩾ 0.5 D.) Ordinate: cumulative percentage; abscissa (degrees): difference [S-O].

ferences [S-O] in cylinder axis components (degree) between I.O.T.A. refractor and subjective refraction is shown in Fig. 11. Ninety percent of the
I.O.T.A. refractor axis measurements differ by 10 degrees or less from those
of the subjective refraction.

7. Binocular Balance

In the measurements given above, each eye was studied separately in the comparisons between data of subjective refraction and those of I.O.T.A. refractor. It may also be interesting to study the discrepancy of refraction between the two eyes of a patient. More precisely, an evaluation of binocular
balance can be obtained by calculating the difference in spherical equivalent refraction between right and left eye, i.e., Δ_{RL}. Eleven subjects out
of the 86 subjects examined have been excluded for various reasons (one-eyed
subjects, or subjects that could not be examined either by subjective methods
or objective refraction).

Among the 75 subjects considered, 70 (93%) had a binocular balance below
2 D, the remaining subjects (7%) showing a difference Δ_{RL} ranging from a
minimum of 2.5 to a maximum of 4 D.

Binocular balance can be calculated by $(\Delta_{RL})_S$ for the subjective refraction and by $(\Delta_{RL})_O$ for the objective refractor; the difference between these
two values is given by the expression:

$$(\Delta_{RL})_S - (\Delta_{RL})_O = S - O .$$

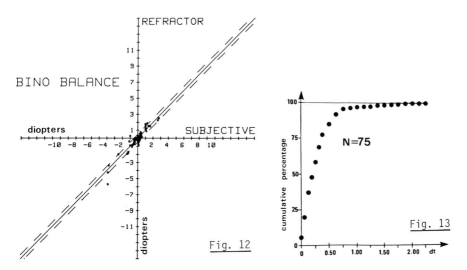

Fig. 12 Binocular balance (in diopters) of I.O.T.A. refractor (ordinate)
and subjective refraction (abscissa) (N = 75 subjects). Coefficient of
correlation r_o = 0.95. Regression line slope p_o = 1.06 and intersects abscissa at y_o = 0.04 dt. Sample standard error of estimate s' = 0.40 dt.

Fig. 13 Cumulative frequency distribution of differences in binocular balance between I.O.T.A. refractor (ordinate) and subjective refraction (ab-
cissa); N = 75. Ordinate: cumulative percentage; abscissa (diopters): difference [S-O].

Figure 12 shows the correlation between I.O.T.A. refractor and subjective refraction binocular balance data $[(\Delta_{RL})_S$ abscissa $(\Delta_{RL})_0$ ordinate].

Figure 13 represents the cumulative frequency distribution of differences [S-O] in binocular balance between I.O.T.A. refractor and subjective refraction. Eighty-five percent of the binocular data are below 0.% D.

This cumulative frequency distribution of differences [S-O] in binocular balance is in agreement with results on monocular data as shown in Figs. 5, 7, 9, and 11.

8. Conclusion

The I.O.T.A. objective refractor makes it possible to see immediately the ophthalmoscopic image of a point on a patient's retina. The microprocessor relieves practitioners of tedious tasks while ensuring the acquisition of data over a very short period. The statistical computations and the various plots show good agreement between measurements given by objective refractor and subjective examination.

Acknowledgments

The work reported in this paper was supported by a contract from the DRET - 78-329. The authors would like to thank P. Fournet and J.C. Rodier for their help in design, setting up, and adjustments of the prototype.

References

1 A. Safir, H. Knoll, and R.C. Mohrman: Trans. Amer. Acad. Ophthal. and Otolaryng, 1266 (Nov.-Dec. 1970)
2 H. Knoll, R.C. Mohrman, W.L. Maier: Amer. J. Optom. and Arch. Amer. Acad. of Optom. 47, 644 (1970)
3 R.C. Mohrman and J.G. Hogan: The Optician, March 11, pp. 20-29 (1977)
4 H.I. McDevitt: The Optician, March 20, pp. 33-42 (1977)
5 P.G. Sloan and K.A. Polse: Amer. J. Optom. and Physiol. Optics, 51, 189 (1974)
6 K.A. Polse, and K.E. Kerr: Arch. Ophthal. 93, 225 (1975)
7 C.R. Munnerlyn: First International Congress Ophthalmological Optics, Tokyo (May 1978)
8 D.L. Guyton: The Optician, June 17, pp. 7-11 (1977)
9 L.S. Sasieni: The Optician, June 20, pp. 12-22 (1980)

Direct Recording of Accommodative Response versus Accomodative Stimulus

Kazuhiko Ukai, Yasuyuki Tanemoto, and Satoshi Ishikawa

Department of Ophthalmology, School of Medicine, Kitasato University, Kanagawa 228, Japan

1. Introduction

Two-dimensional recordings of accommodative response (Ar) versus accommodative stimulus (At) are necessary in order to obtain detailed information of the accommodative mechanism. A new method has been attempted by means of a modified infrared optomer. Using this device, accommodative responses were recorded in normal subjects with emmetropia, myopia and hyperopia, in an amblyopic patient, and in other patients with neurological diseases that disturb accommodation.

2. Method

This study was carried out after modification of a commercially available automated refractometer. The general appearance of the instrument (Nidek model AR-2000) is shown in Fig. 1. A block diagram of its operation is shown in Fig. 2. A target presentation was made by a Badal optical system [1].

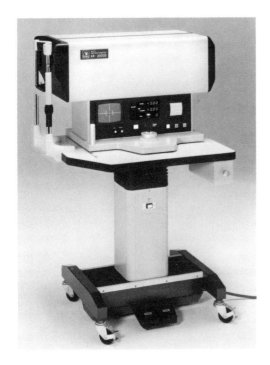

Fig. 1 The automated refractometer (Nidek Ar-2000). An infrared television is used for monitoring the eye. No modification of the external appearance is produced by the internal changes

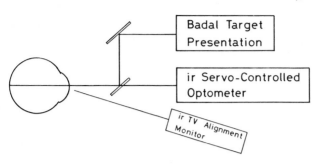

Fig. 2 Schematic of the automated refractometer (ir = infrared).

An infrared (ir) servo-controlled optometer of the Cornsweet [2] type was used to measure the refraction of the eye. Resolution can be made up to 0.005 diopters (D), and an absolute error of the optometer is less than 0.04 D over the entire range. Minimum requirement of the pupil diameter is 2.9 mm. Movements of the eye, blinks, as well as the state of the iris were monitored through infrared television so that malalignment of the equipment could be automatically detected. An autofogging system is employed in this automated refractometer, which is effected by a computer-controlled target system.

The stimulus changes were then made in two ways: 1) using the software sequence which controls the target position, and 2) monitoring the analogue outputs of both target position and the state of refraction. To realize the first control, a dual central processing unit (CPU) system was employed. A master CPU in the AR-2000 controls both optometer and target position as well as the display panel. A slave CPU, which has a small keyboard and with flexible programming, controlled the target position. In the front panel, both stimulus and response are displayed with digitized values. Analogue outputs, the second source of stimulus-response control, were fed from potentiometers in the servo-controllers of both target and optometer. The change from automated refractometer mode to accommodation study mode and vice versa was made with a single switch. However, the expressions of diopter of both stimulus and response in this work were measured at a point 12 mm in front of the subject's cornea vertex, not from the cornea vertex itself.

Direct recording of accommodative response versus accommodative stimulus (Ar-As) was made as shown in Fig. 3. A target was controlled by a micro-computer to elicit smooth accommodative response from -12.5 D to +12.5 D and returning to -12.5 D. The optometer was continually monitoring the changes. The refraction of just the horizontal meridian of the ipsilateral eye was measured in this work. Analogue signals from both the stimulus position and accommodative response were received from the modified optometer. The Ar-As function was recorded on a X-Y recorder and the dynamic response of accommodation against time was recorded on a rectilinear pen-recorder combined with the stimulus of accommodation.

The dimensions of the target are shown in Fig. 4. The illuminated field subtended an angle of 8°, and its luminance was about 10 cd/m^2. The fixation target was located at the center of the field and consisted of a black asterisk subtending an angle of 3°. Each bar was 40 minutes of arc. With the aid of the Badal optical system, no size change existed over the range of accommodation. The surround field except for the target was dark. The refraction was measured under natural pupil size; no mydriatics were used.

Fig. 3

Apparatus

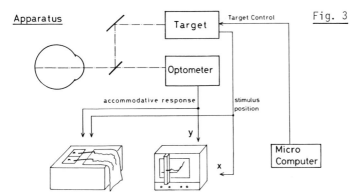

Target Dimensions

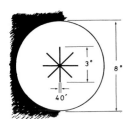

Luminance :

$\approx 10\,cd/m^2$

Contrast :

$\approx 100\,\%$

Fig. 3 Schematic of the control portions of the apparatus for direct recording of accommodative response (Ar) versus accommodative stimulus (As). Control of the target is performed by a microcomputer

Viewing Conditions

monocular
natural pupil

Fig. 4 Dimensions of the target

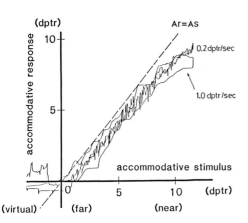

Fig. 5 Hysteresis produced by varying the velocity of the target during measurement of the accommodative response in a normal subject. Two Ar-As curves are presented. One Ar-As curve shows the data obtained with the target velocity at 0.2 D/sec. The enclosed Ar-As loop (indicated by the arrow) shows data obtained with a velocity of 1.0 D/sec and resultant hysteresis.

The velocity of the target was fixed at 0.2 D/sec throughout the study. Figure 5 shows typical results obtained from two different speeds of the accommodative stimulus. When the velocity of 1.0 D/sec is employed as given by an oblique arrow, the hysteresis, i.e. decreased response associated with increasing stimulus and increased response with decreasing stimulus, increased during the measurement. But, when 0.2 D/sec is used, the curve shows less hysteresis and good reproducibility is obtained. Almost all normal subjects show no hysteresis using the target velocity of 0.2 D/sec. In the Ar-As graph, accommodative stimulus is given on the abscissa and accommodative response is given on the ordinate, both expressed in diopters.

3. Results and Discussions

3.1 Emmetropia and Ammetropia

A superimposed curve of Ar-As recording of both increasing and decreasing accommodative stimuli obtained from 5 normal subjects who were approximately emmetropic is given in Fig. 6. The age ranged from 22 to 30 years old with a mean age of 25. The dotted line in the first quadrant indicates the theoretical line on which the accommodative response coincides with accommodative stimulus. The actual data fall on a line slightly less steep than the theoretical line. The far point of accommodation is located approximately at 0 D, and the near point of accommodation is at about 7 D, therefore the amplitude of accommodation is also 7 D. There are two groups of subjects categorized as to their near point of accommodation; namely, one group who could keep accommodating at the near point and a second group who could not hold accommodation at that point. The latter are notated on the right side of the figure as the set of curves that dip sharply downward.

The negative area in Fig. 6 means that the target is virtual image, equivalent to the strength (load) of the positive lens. Even at high values of positive lens loading shown in the second quadrant, two subjects seem able to accommodate. This may be due to forced accommodative effort under the viewing condition and to instrument myopia. The other three subjects revealed a refraction of less than 1 D at this condition. Ar-As recordings thus obtained from myopia of -4.0 D (Fig. 7, upper recording) and hyperopia of +2.0 D (Fig. 7, lower recording) are shown. Their accommodative far point is clearly different from that of emmetropia. The slopes are almost equal to that of emmetropic subjects but the lags which will be described seem to be larger than the emmetrope, especially in the hyperopic eye.

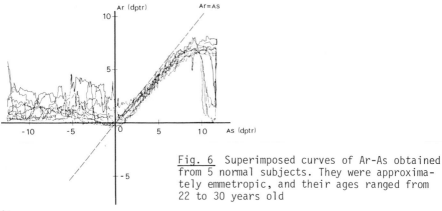

Fig. 6 Superimposed curves of Ar-As obtained from 5 normal subjects. They were approximately emmetropic, and their ages ranged from 22 to 30 years old

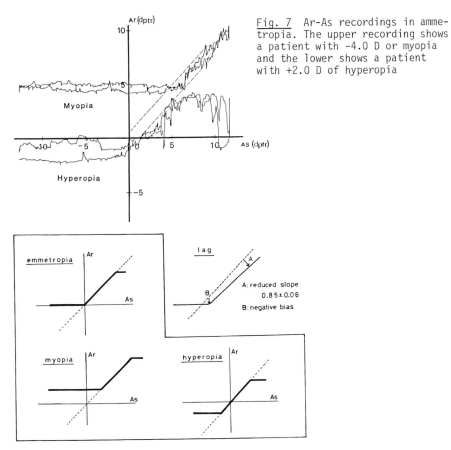

Fig. 7 Ar-As recordings in ammetropia. The upper recording shows
a patient with -4.0 D or myopia
and the lower shows a patient
with +2.0 D of hyperopia

Fig. 8 Schematic drawings of Ar-As curves for various refractive states.
The most remarkable difference of the Ar-As curve between emmetropia and ammetropia is the shift of accommodative far point. The accommodative near point
also shifts. These shifts are influenced by the amplitude of accommodation.
Two components of accommodation lag, A: reduction of the slope, and B: negative bias, respectively, are illustrated in the upper right corner.

Schematic drawings of the Ar-As curve under the various conditions of
refraction are shown in Fig. 8. The upper left shows the emmetropia and the
lower two figures show the ammetropia. These figures are idealized, but in
the actual recordings, accommodation lags exist as shown in Figs. 6,7. The
lags are analyzed as the reduced slope and the negative bias as shown in the
upper right of Fig. 8. The mean slope of 30 normal subjects (age range: 20
to 30) is 0.85 with the standard deviation of 0.06. The mean negative bias
is less than about a half diopter in emmetropia but was increased in amme-
tropic eyes.

3.2 Amblyopia

Figure 9 shows the Ar-As recording from a monocularly amblyopic patient produced by visual deprivation since he had a one-sided congenital ptosis of

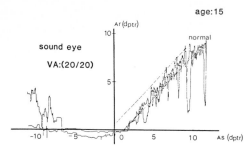

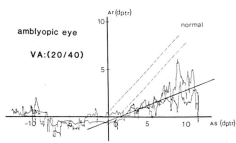

Fig. 9 Ar-As recording obtained from a monocular amblyope. Corrected visual acuities of his sound and amblyopic eye were 20/20 and 20/40, respectively

mild degree. These figures were recorded from his normal eye (upper) and his amblyopic eye (lower) whose corrected visual acuity was 20/40. An obvious reduction of the slope with good repeatability can be seen in the amblyopic eye with reduced amplitude of accommodation. The normal range of accommodation in this age group is given by two dashed lines.

The results obtained from anisometropic and strabismic amblyopia are fundamentally the same, and coincide well with the static accommodation study using a laser speckle optometer in amblyopia carried out by WOOD and TOMLINSON [3]. The reason why the ratio of accommodative response to the accommodative stimulus (Ar/As ratio: slope of the Ar-As recording) is decreased can be explained by the following analogy. In normal subjects, the viewing conditions such as decreased luminance of the target [4], targets composed of sinusoidal gratings with low spatial frequencies [5] or decreased pupil diameter [6], produce a reduction in the Ar/As ratio. In these cases, defocusing of the eye causes less effect on the perceived image because of reduced visual acuity and less contrast reduction as compared with higher frequency gratings or increased depth of focus. Thus, less accuracy of accommodation is required in such situations.

3.3 Neurological Diseases

Ar-As recordings in patients with two neurological diseases are presented. Figure 10 shows a recording from an 18-year-old patient with Fisher's syndrome. The onset of the disease was 12 May, with general neurological symptoms of areflexia. Almost complete palsy of extra ocular muscles was observed, and slightly tonic pupils were seen. During the process of recovery (3 weeks after the onset), Ar-As was recorded. At this time, a reduced amplitude of accommodation in bilateral eye movement was observed (upper figure). Eye movement was slightly improved but still paralyzed. Two days later (on 2 June) Ar-As was recorded again. The amplitude of accommodation was well-recovered at this time. On 15 June, general neurological signs were almost gone, and eye movement recovered completely. The reduced amplitude of ac-

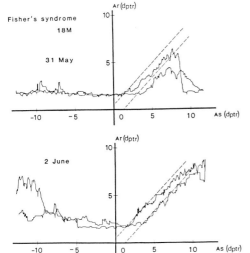

Fig. 10 Ar-As curves in a patient with Fisher's syndrome. Ar-As records were obtained during the course of a three-day recovery period. The reduced amplitude of accommodation returned to the normal range (----) on the third day of measurement

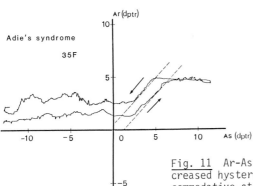

Fig. 11 Ar-As curves of Adie's syndrome. Increased hysteresis is observed when the accommodative stimulus is increasing (ascending arrow) and also decreasing (descending arrow)

commodation may be due to peripheral paresis of accommodation. This case seemed to be a typical Fisher's syndrome [7] as judged by the clinical course.

Figure 11 shows the recording from a 35-year-old patient with Adie's syndrome. Abnormal accommodation tonicity associated with Adie's syndrome has been described, however no quantitative study has been made. An obvious hysteresis can be seen both at increasing and decreasing accommodative stimuli. In this case, the target velocity of 0.2 D/sec appeared to be too fast. Note that this hysteresis also appeared in normal subjects when the target velocity of accommodation in Adie's syndrome is definitely reduced and prolonged.

In order to confirm this suspicion, the dynamic response of the accommodation by step displacement of the stimulus was measured. The result is given in Fig. 12 (lower) with a normal subject (upper) for comparison. As the stimulus changed from 2 D to 4 D in stepwise fashion, the accommodative

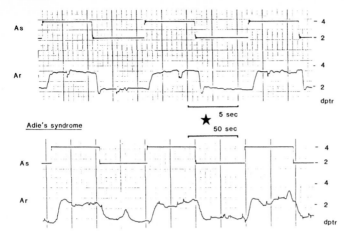

Normal

As

Ar

dptr

5 sec
50 sec

Adie's syndrome

As

Ar

dptr

<u>Fig. 12</u> Dynamic responses of accommodation produced by a step stimulus (As)
from 2 to 4 D. The (Ar) responses of a normal subject (upper) and for a
patient with Adie's syndrome (lower) are given. The lower figure was obtained from the same patient as in Fig. 11. Note that the time scale is ten
times slower for the Adie's syndrome patient.

response can be seen in the second and fourth recordings. When calculated,
the velocity of accommodation is about 3.0 D/sec in the normal subject and
about 0.3 D/sec in the case of Adie's syndrome. One should note that the
time scale in the recording of Adie's syndrome is 10 times slower than that
of the normal subject.

4. Conclusion

With modification of existing equipment, one can perform quantitative measurements of the accommodative system. Using this method, normal accommodative
function can be examined in detail in a quantitative manner, as well as various functional diseases of the eye including patients with various neurological diseases with symptoms of accommodation difficulty.

References

1   P. Badal: Ann. d'Oculist <u>75</u>, 101 (1876)
2   T.N. Cornsweet and H.D. Crane: J. Opt. Soc. Amer. <u>60</u>, 548 (1970)
3   I.C.J. Wood and A. Tomlinson: Amer. J. Optom. Physiol. Opt. <u>52</u>, 243 (1975)
4   C.A. Johnson: J. Opt. Soc. Amer. <u>66</u>, 138 (1976)
5   W.N. Charman and J. Tucker: Amer. J. Optom. Physiol. Opt. <u>55</u>, 84 (1978)
6   R.T. Hennessy et al.: Vision Res. <u>16</u>, 587 (1976)
7   F.B. Walsh and W.F. Hoyt: *Clinical Neuro-ophthalmology*, 3rd ed., Vol. 2
    (The Williams and Wilkins Co., Baltimore, 1969), pp. 1026-1241
8   F.B. Walsh and W.F. Hoyt: ibid., Vol. 1, pp. 434-566

Lag of Accommodation

Mototsugu Saishin, Hiroshi Uozato, Hiroyuki Makino, and Shuitsu Nakao
Department of Ophthalmology, Nara Medical University
Kashihara-shi, Nara-ken, 634, Japan

Kimihiro Yamamoto
Health Administration Center, Nara Women's University, Nara-shi, 630 Japan

Introduction

There are many concepts of lag of accommodation [1,2]. In this paper it is defined as follows: regardless of how clear an image an accommodating eye can maintain of a nearer object, its retinal conjugate point does not reach that object (Fig. 1).

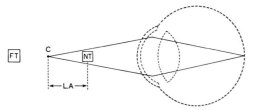

FT: far target
NT: near target
C : retinal conjugate point
LA: lag of accommodation
    (Diopter)

Fig. 1 Lag of accommodation occurs when an eye shifts its regard from a far target to a target between the far one and the eye's punctum proximum. In this situation the retinal conjugate point is beyond the nearer target regardless of the eye's effort to maintain distinct vision.

We determined the frequency and amount of the lag of accommodation in young healthy women under natural visual circumstances by using an objective autorefractometer. The incidence was about 100% in the ocular second principal meridian and its mean value was 1.1 D when the subject's eye maintained distinct vision on the target presented at 0.5 m in front of the eye.

The occurrence of lag of accommodation may call into question the validity of clinical examinations concerning accommodation such as punctum proximum, amplitude of accommodation, AC/A ratio, etc.

Methods

Subjects were 18-20 year old students in their freshman year at a women's university. They were divided into four groups at random and some with one or both eyes more myopic than -2 D in the first principal meridian were excluded from the analysis in this paper. The groups and eye conditions of the subjects are listed in Table 1.

The refractions of all subjects in group I were measured with fixation of a Landolt ring of decimal visual acuity value 0.1 shown at a 5 m distance.

Table 1 Number of subjects in each category

group	condition of eye when measured / original naked eye	without correction	with HCL	with SCL	
I	28	5	16	9	58
II	7	1	10	6	24
III	13	5	7	3	28
IV	19	5	5	10	39
total	67	16	38	28	149

Second, 34 members of group I were measured with a Landolt ring of 0.06, and 24 members with a ring of 0.1 shown at a 0.5 m distance in a well-lit room (50 lux). The direction of the gap of the Landolt ring was upward throughout this experiment.

In measuring group II, the refraction at 5 m distance was achieved under the same condition as with group I; then the room lights were dimmed to 1-3 lux and a Landolt ring of size 0.1 moved to 0.5 m for a near measurement.

The experiments with groups III and IV were achieved with the same light conditions in the room as for groups I and II, respectively. The only difference was the use of a dimmed pocket light that was usually used in clinics as the fixation target instead of a Landolt ring [3,4].

The fixation targets were set up to move along the alignment of the optical axis of the refractometer and the visual line of the subject's right eye as shown in Fig. 2. Consequently the measurements were achieved with binocular vision and the data for analysis in this paper was obtained entirely from the performance of right eyes of the subjects [5,6],

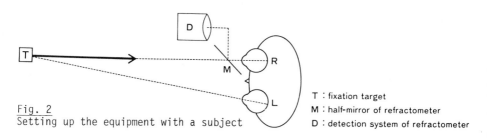

Fig. 2
Setting up the equipment with a subject

T : fixation target
M : half-mirror of refractometer
D : detection system of refractometer

The refraction was estimated by an automatic objective refractometer, Canon R-1 [7]. After the instrument's optics were aligned with the subject's eye on the refractometer's TV monitor screen, a button was pressed and the sphere, cylinder and cylinder axis values were given. At the same time the TV screen, which was divided into two sections by a mixing amplifier, displayed the fixation target and its position. All information on the TV screen was recorded by a video recorder.

The refraction was estimated three times for each condition. If a subject's eye was out of alignment with the refractometer during the measurement, the estimation was continued until completed successfully three times.

Results

Referring to Fig. 3, let T be the position of the fixation target, 0.5 m in front of a subject's eye expressed dioptrically. And let W, E, and S be the positions, expressed dioptrically, of the conjugate point in the second principal meridian, the circle of least confusion, and the conjugate point in the first principal meridian, respectively. The values of W, E, and S of each subject were calculated from values given by the refractometer. The conjugate point of the eye without astigmatism is defined as being at the circle of least confusion.

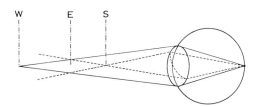

W : the point in 2nd principal meridian
E : the point of the circle of least confusion
S : the point in 1st principal meridian

Fig. 3 Retinal conjugate point

The subjects were categorized according to the relative positions of the target and their conjugate points (Table 2).

If in three successive measurements a subject's W, E, or S conjugate point (but not more than one of the three conjugate points) equalled T one or more times with errors remaining less than ±0.12 D, the subject was categorized in the corresponding cell. Next, those subjects whose conjugate points never equalled T and whose conjugate points fell into one of the remaining categories all three times were categorized accordingly. Six subjects remained to be categorized. One subject from group I gave readings of T=E, T=S and W>T>E was counted in the group I T=E cell. Three subjects in group I gave readings of T=S for all three measurements but for two the error was less than ±0.25 D and for the other less than ±0.50 D. They were counted in the group I T=S cell. One subject in group III and another in group IV also gave readings of T=S with errors of less than ±0.25 D. They were counted in their groups' T=S cell. Two subjects whose readings defied categorization were excluded from this paper.

The percentage of the subjects from each group which fell into the different T-related categories is given in Table 3. Table 4 shows the percentage of occurrence of lag of accommodation in the first principal meridian, at the circle of least confusion, and in the second principal meridian.

Table 2 Categorization of subjects according to relative positions of target and conjugate points. The conjugate point of the eye without astigmatism is defined as being at the circle of least confusion.

group: condition when measured	T>W	T=W	W>T>E	T=E	E>T>S	T=S	S>T	total
I : visual chart in light	0	0	0	9	2	34	11	56
II : visual chart in dark	0	0	1	3	1	9	10	24
III : pen-light in light	0	0	0	7	0	16	5	28
IV : pen-light in dark	1	0	3	6	4	12	13	39
total	1	0	4	25	7	71	39	147

71

Table 3 Categorization of subjects expressed as percentages

group: condition when measured \ position of conjugate point relative to target	T>W	T=W	W>T>E	T=E	E>T>S	T=S	S>T	total
I : visual chart in light	0	0	0	16	4	61	20	100
II : visual chart in dark	0	0	4	13	4	38	42	100
III : pen-light in light	0	0	0	25	0	57	18	100
IV : pen-light in dark	3	0	8	15	10	31	33	100

Table 4 Frequencies of lag of accommodation

group: condition when measured at 0.5m \ ocular meridian of conjugate point	in the 1st principal meridian	at the circle of least confusion	in the 2nd principal meridian
I : visual chart in light	20	84	100
II : visual chart in dark	42	83	100
III : pen-light in light	18	75	100
IV : pen-light in dark	33	74	97
total	27	79	99.6

Table 5 Mean value of lag of accommodation (Diopter)-point estimation

group: condition when measured at 0.5m \ ocular meridian of conjugate point	in the 1st principal meridian	at the circle of least confusion	in the 2nd principal meridian
I : visual chart in light	0.1	0.6	1.1
II : visual chart in dark	0.2	0.8	1.3
III : pen-light in light	−0.2	0.3	0.8
IV : pen-light in dark	0	0.6	1.1
mean	0	0.6	1.1

Table 6 Mean value of lag of accommodation (Diopter)-interval estimation with confidence coefficient of 95%

group: condition when measured at 0.5m \ ocular meridian of conjugate point	in the 1st principal meridian	at the circle of least confusion	in the 2nd principal meridian
I : visual chart in light	−0.1~0.3	0.4~0.8	0.9~1.3
II : visual chart in dark	−0.1~0.5	0.5~1.1	1.0~1.6
III : pen-light in light	−0.5~0	0~0.5	0.5~1.1
IV : pen-light in dark	−0.2~0.3	0.4~0.8	0.9~1.3

Table 5 shows the mean value of the lag of accommodation of eyes fixating on the target presented at 0.5 m. Table 6 shows the same values expressed as interval estimation with a confidence coefficient of 95%.

Discussion

Sheard (1922) recognized the phenomenon of lag of accommodation by using dynamic retinoscopy [1,2]. Since then many investigators have studied the phenomenon. According to BORISH [2], lag of accommodation has meant different things to different investigators, basically because of variations in the technique of dynamic retinoscopy. Furthermore, all the investigators worked in a dark room which is different from natural visual circumstances.

Recently several studies concerning lag of accommodation were conducted by using infrared optometers that were designed originally to measure refractive change in only one ocular meridian plane which is determined by the optics of the optometer [e.g. 8]. They could not get quantitative refraction values of the accommodating eye in all meridians.

From the information obtained in these ways lag of accommodation was considered to be of little use in evaluating clinical data on accommodation.

Our results show that the frequency of lag of accommodation is about 100% in the second principal meridian, 80% at the circle of least confusion and 30% in the first principal meridian in young healthy women. The mean amount of lag of accommodation was 1.1 D in the second principal meridian, 0.6 D at the circle of least confusion, when the fixed target was 0.5 m in front of the eye.

On the other hand the validity of calibration between visible and infrared light employed in the refractometer was ensured by arbitrarily picked subjects whose refraction was estimated by the usual subjective examinations. The refractometer used in this study was a commercially available model [7]. In addition, according to our observation of the measurement records, viewing the target at 0.5 m did not constrict the pupil [9]. This may deny the possibility that the lag of accommodation is due to increased depth of focus.

Conclusion

Therefore it may be concluded that lag of accommodation occurs constantly in the second principal meridian.

If, for example, a clinician is gauging the punctum proximum he can make a serious error because without taking lag of accommodation into consideration he will determine the accommodating refraction to be -2 D when the fixed target is 0.5 m from the eye. In actuality the accommodating eye's refraction will be 0.9 D in the second principal meridian.

Likewise, estimations concerned with accommodation, such as punctum proximum, amplitude of accommodation, AC/A ratio [10], etc., that do not account for lag of accommodation may be unreliable.

References

1   S. Duke-Elder and D. Abrams: In *System of Ophthalmology*, Vol. 5, ed. by
    S. Duke-Elder (Henry Kimpton, London, 1970), p. 451
2   I.M. Borish: *Clinical Refraction* (Professional Press, Inc., Chicago, 1975), p. 697

3   M. Alpern: J. Opt. Soc. Amer. $\underline{48}$, 193 (1958)
4   F.W. Campbell and G. Westheimer: J. Opt. Soc. Am. $\underline{49}$, 568 (1959)
5   R. Rosenberg, N. Flax, B. Brodsky, and L. Abelman: $\overline{Am}$. J. Optom. Physiol. Opt. $\underline{30}$, 244 (1953)
6   J. Semmelow and N. Venkiteswaran: Vision Res. $\underline{16}$, 403 (1976)
7   I. Matsumura, S. Maruyama, Y. Ishikawa, R. Hirano, Y. Kobayashi, and Y. Kobayagawa: Abstract of 16th Ophthalmological Optics of Japan $\underline{2}$ (1980)
8   F.W. Campbell and J.G. Robson: J. Opt. Soc. Amer. $\underline{49}$, 268 (1959)
9   F.W. Campbell: Opt. Acta. $\underline{4}$, 157 (1957)
10   D.D. Michaels: *Visual Optics and Refraction, a Clinical Approach* (C.V. Mosby, St. Louis, 1980)

The Effects of Various Eye Diseases on the Measurement of Refractive Error Using the Nikon Autorefractometer

Akio Hosaka and Hiroshi Tagawa

Department of Ophthalmology, Asahikawa Medical College, 4-5 Nishikagura, Asahikawa, Japan

Kazuo Morohashi

Nippon Kogaku, K.K., Tokyo, Japan

Introduction

HOSAKA et al. [1] previously reported the correlation study of the Nikon NR-1000 Refractometer (hereafter referred to as NR) and subjective testing in evaluating refractive error. In the present study, the validity of the instrument for measuring refractive error in pathologic eyes was studied.

Subjects and Methods

The subjects were taken at random from first-time patients between December 1980 and June 1981, and also between October 1981 and March 1982, at The Eye Clinic, Asahikawa Medical College. In total, 227 cases (349 eyes) ranging from 5 to 87 years of age were examined (Table 1 and Table 2).

Table 1 Number of cases by age and sex group

age \ sex	~ 9	10 ~	20 ~	30 ~	40 ~	50 ~	60 ~	70 ~	(total)
male	16	14	7	9	11	14	21	17	109
female	12	7	7	10	17	16	27	22	118
(total)	28	21	14	19	28	30	48	39	227

Measurements with the NR were made at least four times in each eye, and the readings with highest confidence value were employed. Occasionally, the Hartinger's refractometer test was added as an additional, precise, subjective test. The NR is controlled electronically and refractive readings are printed out. Furthermore, NR provides a confidence factor value analyzed by a microcomputer. Thus, 90 or higher, between 80 and 89, and less than 80 are called, respectively, good (A), fair (B), and poor (C), respectively. In some cases, unexpected refractive values (D), mostly strong plus power, were detected and in others, the measurement was impossible (E), because of the pathologic conditions of the eyes.

Table 2 Number of eyes by disease group

	male	female	(total)
cataract	48	80	128
amblyopia	35	39	74
chorioretinopathy	32	22	54
maculopathy	8	12	20
optic neuropathy	10	8	18
corneal disease	16	10	26
iritis	3	5	8
glaucoma	2	5	7
retinal detachment	8	3	11
vitreous opacity	1	2	3
(total)	163	186	349

Table 3 Relation between vision and NR measurement (Group 1)

	vision	total number of eyes	fair or poor (B, C)	unexpected (D)	impossible (E)	either D or E
cataract	> 0.4	63	9 (14.3%)	1 (1.6%)	5 (7.9%)	6 (9.5%)
cataract	0.1-0.3	37	5 (13.5%)	5 (13.5%)	18 (48.6%)	20 (54.1%)
cataract	< 0.1	28	12 (42.9%)	19 (67.9%)	10 (35.7%)	24 (85.7%)
corneal disease	> 0.4	12	5 (41.7%)		1 (8.3%)	
corneal disease	0.1-0.3	6			6 (100.0%)	
corneal disease	< 0.1	8	2 (25.0%)	2 (25.0%)	3 (37.5%)	
vit. opa.	> 0.4	1				
vit. opa.	< 0.1	2		2 (100.0%)		
iritis	> 0.4	8	1 (25.0%)		4 (50.0%)	

Results

Cataract: In viewing Table 3 (upper), the number of D and/or E increases
with lower vision. However, excluding E, in 58 eyes of which vision is 0.4
or better (Table 5), the NR values correlate with the subjective values. The
rate of correlation does not differ so much from that of the normal subjects.

   Corneal disease (Tables 3 and 5): With eyes where vision is 0.3 or worse,
11 out of 14 (78.6%) belong to D or E. With eyes of 0.4 or better, 11 out of
12 cases, the NR values correlate with the subjective values. In all 10 eyes
of E, pathological changes cover the pupillary area.

   Iritis (Table 3): Four eyes belong to E and one eye to C, although the
vision of all 8 eyes is better than 0.4. It suggests the disturbance of incident ray passage due to inflammatory conditions of the eye.

   Vitreous opacity (Table 3): As with iritis, the measurement results depended on the degree of opacity.

   Chorioretinopathy (Tables 4 and 5): In 27 eyes where vision is 0.4 or
better, the correlation between the subjective and NR values is similar to
that of normal subjects. NR values do not differ so much from the subjective
values even in the cases whose vision is 0.1 - 0.3.

Table 4   Relation between vision and NR measurement (Group 2)

	vision	total number of eyes	fair or poor (B, C)	impossible (E)
chorioretinopathy	> 0.4	27	4 (14.8%)	
chorioretinopathy	0.1 - 0.3	6	3 (50.0%)	1 (16.7%)
chorioretinopathy	< 0.1	21	6 (28.6%)	8 (38.1%)
maculopathy	> 0.4	8	2 (25.0%)	
maculopathy	0.1 - 0.3	6		1 (16.7%)
maculopathy	< 0.1	6		2 (33.3%)
ret. detachment	> 0.4	5		2 (40.0%)
ret. detachment	0.1 - 0.3			
ret. detachment	< 0.1	6	1 (17.0%)	3 (50.0%)

Table 5 Correlation between Subjective Refraction and Nikon NR-1000 Refractometer*

	sphere ± 0.5D	cylinder ± 0.5D	cyl. axis ± 10°
cataract	44/58	40/58	33/38
	75.0%	69.0%	86.8%
corneal disease	9/11	7/11	6/8
	81.8%	63.6%	75.0%
chorioretinopathy	20/27	16/27	13/14
	74.0%	59.0%	93.0%
maculopathy	5/8	7/8	4/5
	62.5%	87.5%	80.0%
amblyopia	26/40	26/40	22/29
	65.0%	65.0%	75.9%

* eyes of vision 0.4 or better only

Maculopathy: Only 5 eyes out of 20 showed B and E (Table 4). These 5 eyes show moderate or intense edema or degeneration of the macula. In 8 eyes of which vision is 0.4 or better, the correlation between the subjective and NR values is not too bad.

Retinal detachment (Table 4): Any direct correlation between the confidence value and the vision could not be determined.

Amblyopia (Table 6): Seventeen eyes out of 73 (23%), one eye with poor fixation excluded, had measurements in the B or C categories. Among 17 eyes, astigmatism of 1.5 D or more in 15 eyes (88.2%), and of 3.0 D or more in 13 eyes (76.5%) was detected. In all 74 eyes, astigmatism of 2.0 D or more was found in 29 eyes (39.2%). Of 40 eyes where vision is 0.4 or better, the correlation between the subjective and NR values is slightly low.

Optic neuropathy: Reliable, A category measurements were made in 16 out of 18 eyes. The other 2 belong to E, because of poor fixation.

Glaucoma: Reliable, A measurements were made in 5 out of 7 eyes. The other 2 belonged to E, because of poor fixation.

Discussion

It has been approximately 10 years since automated refractometers were first introduced [2]. These instruments have become one of the effective measures for refractions [3]. Most of the previous papers deal only with the accuracy of measuring the refractive error.

PAPPAS et al. [4] pointed out the possible sources of error in Acuity System's Autorefractor 6600--poor fixation, high myopia and opacity in the media. BIZZELL [5] discussed the diseases which reduce the transparency of the media as a theoretical limit of the Bausch & Lomb Ophthalmometron.

The source of light for measuring refractive error in the NR has a peak wavelength of 880 nm and is completely invisible. The refractive error is measured twice over 180 degrees, at the interval of 1 degree, namely 360 readings in each eye (time required, 0.5 sec). The results are calculated by means of a microcomputer [6]. In the present study, employing the common cases of eye disease, we studied the limits of measurement and usefulness of the Nikon Autorefractometer.

In the cataract cases, the measurement was impossible in most eyes with acuities under 0.1 and in more than half the eyes with acuities between 0.1 and 0.3. However, in most of the eyes of 0.4 or better, excluding 6 eyes (9.5%), the measurement was possible with nearly the same accuracy as with normal subjects. The results seemed to be due to the predominance of senile cataract in which the opacity progresses usually from the periphery to the center of the lens. In corneal diseases, similar results were found suggesting that NR measurement as well as vision depends on the condition of the central portion of the eye. Also, in most of the eyes of 0.4 acuity or better, the results of measurement showed nearly the same accuracy as in the normal subjects. In the cases of iritis, the measurement was available only in 3 eyes out of 8, in spite of relatively good vision. It seemed probable that inflammatory conditions may disturb or scatter the incident light. In the cases of vitreous opacity, the situation may be the same. The common element in the above 4 groups is that the incident ray may be disturbed before reaching the fundus due to the opacity of the media. It is characteristic in this group that unexpectedly high plus values were detected, especially in the highly opaque eyes. It is probable that the source of light might be reflected at the surface of the opacity.

In the cases of chorioretinopathy, 27 eyes of 0.4 acuity or better showed results similar to those in the normal subjects. Relatively small differences between NR and subjective values, even in eyes showing acuities of 0.1-0.3, may be useful. In the cases of maculopathy, only 8 eyes out of 20 showed vision of 0.4 or better. NR and subjective values in these eyes are correlated with each other. The other eyes showed considerable variations in measurement, which related to the extent and degree of the affected region of the macula. The NR measurement in the cases of retinal detachment also showed considerable variations. Reasonable values were not available where the maculae are involved in the detachment or in secondary changes. The common element in the 3 groups is that the source of light might be abnormally reflected by the fundus, particularly in the cases where the macula is involved. However, the measurement can be made in most cases when the vision was 0.4 or better.

Only one out of 74 amblyopic eyes could not be measured by NR, because of poor fixation. It was remarkable that 17 eyes (23%) showed a low score (B and C) as shown in Table 6. However, it was proved that the result was not due to the characteristic of amblyopia but to an accompanying astigmatism. This is because the confidence value has been set up to be lower with increasing divergence from the theoretical curve ($Sin^2 \theta$), and generally, strong astigmatism is inclined to deviate from the curve. Reliable A measurements were made in 16 out of 18 eyes of optic neuropathy and in 5 out of 7 eyes with glaucoma. The other 3 showed poor fixation. Proper NR measuring is obtainable in these diseases, provided that the source of light falls

Table 6 Confidence value in amblyopia

confidence value	number of eyes
good ≧ 90	56
fair 80 ~ 89	10 ⎫
poor 70 ~ 79	7 ⎬ (23%)
impossible < 70	1

on the macula. The common element in the above 3 groups is that the pathology exists in the central nervous system inclusive of the optic nerve.
Therefore, NR measurement can be possible irrespective of vision unless
other conditions are involved.

By means of testing in 349 eyes with various diseases, wider application
of the NR has been confirmed. Generally speaking, NR measurements, similar
in degree to those of the normal subjects, can be made if the vision is 0.4
or better. In clinical application, pathologic eyes were favorably divided
into 3 categories, each with its individual characteristics.

Summary and Conclusions

We investigated the effects of various eye diseases and the measurement of
refractive error using the Nikon NR-1000 Autorefractometer. The cases were
selected at random among first-time patients. The subjects ranged from 5 to
87 years of age, and totaled 227 (349 eyes). We found:

1. The measurement was insufficient or was prevented in the conditions
where the passage of the incident (infrared) ray through the media was disturbed or irregularly reflected.

2. Pathological conditions involving the macular area, such as edema,
inflammation, or hemorrhage, tended to affect the measurement because of abnormal reflection from the retina.

3. Usually, in the conditions where the incident ray was not disturbed,
a measurement could be made unless other elements secondarily affected fixation, e.g., amblyopia and optic neuropathy.

4. Unexpected strong plus dioptric values were detected in advanced
cataract, opaque cornea and severe vitreous disorders. In those conditions,
the incident ray might be strongly or completely reflected.

References

1   A. Hosaka, et al.: J. Ophthalm. Opt. Soc. Jap. $\underline{3}$, 11 (1982)
2   T.N. Cornsweet and H.D. Crane: J. Opt. Soc. Amer. $\underline{60}$, 548 (1970)
3   S. Ohshima: Jap. Review Clin. Ophthalm. $\underline{76}$, 291 (1982)
4   C.J. Pappas et al.: Arch. Ophthalm. $\underline{96}$, 993 (1978)
5   J.W. Bizzell et al.: Arch. Ophthalm. $\underline{92}$, 102 (1974)
6   M. Nohda et al.: J. Ophthalm. Opt. Soc. Jap. $\underline{1}$, 23 (1980)

Part 3

Applied Optics

A Distortion-Free Cross-Cylinder System for Automated Subjective Refraction

David. L. Guyton

The Wilmer Institute B1-35, The Johns Hopkins Hospital,
Baltimore, MD 21205, USA

Introduction

For the past fourteen years I have been developing instruments for automated clinical refraction. In this paper I should like to describe a major break-through in this work--the development of a distortion-free cross-cylinder system--a system which actually appears to be more accurate than the conven= tional cross cylinder.

In 1887, Edward Jackson described the use of a fixed-power Stokes lens in refining the cylinder power correction in clinical refraction. Twenty years later he described the use of the same lens in refining cylinder axis. Jackson wrote that this lens had become far more useful, and far more used, than any other one lens in his trial set.

Since that time, the Jackson cross cylinder has become perhaps the most important tool in clinical refraction. The cross-cylinder method of refin-ing cylinder axis and power has become the world's standard for accuracy-- the standard against which all other methods are compared.

The cross cylinder was quickly incorporated into refractors in the 1930's in a convenient form. The cross cylinder is popular because it is easy to use, both for refractionist and patient. It is accurate because it refracts all meridians at once, as opposed to the line methods of refraction which are heavily weighted toward the meridians perpendicular to the line targets being used. This is the supreme advantage of the cross cylinder--it refracts all meridians simultaneously.

And yet, the Jackson cross cylinder has limitations--namely the problems of sequential viewing, distortion, and contextural bias. First, the flip choices are viewed sequentially rather than simultaneously. Every refrac-tionist is all too familiar with the patient who always chooses the second choice, or asks repeatedly to see the choices again. The patient himself is left with uncertainty because of lack of opportunity to compare the two choices side-by-side.

Even if the two choices could be compared side-by-side, the second prob-lem would become apparent--distortion. Cylindrical lenses cause distortion, and the cross cylinder is no exception. Mounted 6 to 7 cm in front of the eye, as on a standard refractor, a ±0.25 D cross cylinder produces distor-tion between the two flip choices of 6 to 7%. Even with the best focus balance between the two choices, the patient will often prefer the choice which is vertically elongated. This artifact is well known in cross-cylin-der testing.

The third problem is that of contextural bias. Friedman and Williamson-Noble pointed out in the early 1940's the built-in bias in the letters we use for testing. If one flip choice of the cross cylinder makes the vertical strokes clear, and the other flip choice makes the horizontal strokes clear, the patient will choose the one with the vertical strokes clear because more of the letters are recognizable. This results in giving the patient astigmatism he does not have, or provides him with the wrong correction. Some clinicians use only the letter "O" to avoid this bias, but cross cylinder testing becomes less sensitive when angular detail is missing from the test targets.

Given these limitations of conventional cross-cylinder testing, how might we design a new cross cylinder test--without these limitations--a new test with even greater accuracy than that which we regard as the standard?

Methods

With special optics we can certainly replace sequential viewing with simultaneous side-by-side viewing. This has been done before. Several cross-cylinder systems have been devised in the past twenty years. Some [2] have used prisms, or mirrors and beamsplitters, to provide a double image of a distant target, with each of the double images being viewed through one cross-cylinder choice. These diplopia-principle cross-cylinder systems often suffered from unequal brightness of the two images, difficulty in using black figures on white backgrounds as targets, or problems with maintaining alignment of the patient's eye. In contrast to these previous systems, a new design termed the Simulcross™ cross-cylinder system provides two identical side-by-side targets in a split-field arrangement rather than a double image of a single target. Each half of the split field is viewed through a cylindrical lens mounted in front of that half field to provide appropriate cross-cylinder effect (see Fig. 1).

The Simulcross cross-cylinder system has been incorporated into the AO SR-IV™ programmed Subjective Refractor [3] (Fig. 2), an instrument in which

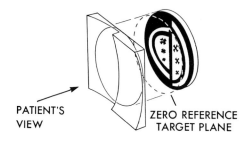

PATIENT'S
VIEW

ZERO REFERENCE
TARGET PLANE

Fig. 1 Simulcross target assembly. Each half of the split-field target has a high-powered plano-cylindrical lens mounted in front of it, and rotating with it, to provide the cylindrical component of the desired cross-cylinder effect (-6.00 D cylinders are used for the ±0.25 cross-cylinder assemblies). The target transparency itself is mounted posterior to the zero reference target plane to provide the spherical component of the cross-cylinders. Each half target is drawn with distortion exactly opposite to that produced by the respective cylindrical lens in front of it, with resulting nullification of distortion in the viewed image. Maltese crosses with sizes corresponding to 20/120, 20/70 and 20/40 letters are viewed on each half target, oriented specifically for minimal contextural bias. The crosses to be compared are separated from each other by approximately 2° between centers.

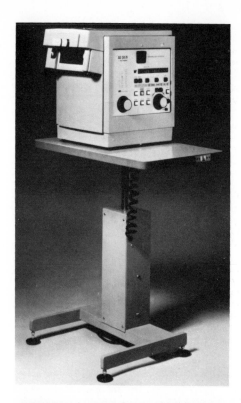

Fig. 2 The AO SR-IV^{TM} programmed subjective refractor

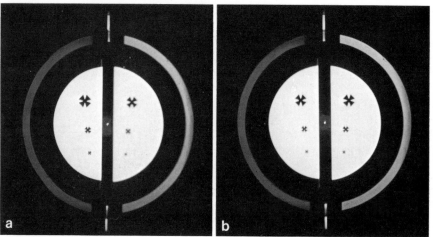

Fig. 3 The Simulcross cross-cylinder target for cylinder power testing, as viewed by the patient. Left: Crosses appear blacker and sharper on the red (left) side. Right: After the operator has pushed the red button, which makes the appropriate change automatically, the crosses appear equal on the two sides, the desired end-point

I have a proprietary interest. The SR-IV is designed to be operated by a technician to provide refined subjective refractions and corrected visual acuities.

The actual distortion produced by each cylindrical lens in the Simulcross assembly is significant, up to 30% with the ±0.50 D assemblies. Previous attempts by HAYNES and by BIESSELS [5,2] at eliminating distortion in cross-cylinder testing have used special lens designs; but with the Simulcross system, distortion is eliminated entirely by drawing the respective half field targets with exactly the opposite distortion to that caused by the cylindrical lenses, such that the resultant images as viewed by the patient are exactly symmetrical in the two half fields (see Figs. 1 and 3) [4].

Maltese crosses are used for the Simulcross targets because of the high sensitivity of acute angles in cross-cylinder testing [1]. If Maltese crosses are oriented properly with respect to the axes of the cross-cylinder through which they are viewed, contextural bias may be minimized [5]. It was confirmed empirically that the arms of the crosses should be 45° to the cross-cylinder axes for least contextural bias, and because the Simulcross targets and lenses rotate as a unit, this ideal orientation can be maintained both for cylinder power testing and for cylinder axis testing.

Thus the Simulcross targets present both cross-cylinder choices side by side, with no distortion and with little or no contextural bias (Fig. 3). The patient judges equality of blur on the two sides of the target, not equality of size or shape. Peripheral red and blue semicircular bands are used to identify the two halves of the target. The patient is asked, "Are the crosses blacker and sharper on the red side, or on the blue side?" And the operator pushes the appropriate colored button according to the patient's response.

There are two pairs of Simulcross assemblies provided, a pair of ±0.25 D target assemblies for routine use, and a pair of ±0.50 D target assemblies for "low vision" refractions. The ±0.50 D Simulcross targets are most useful with visual acuities from 20/40 to 20/70. Whenever the LOW V.A. button is activated, the ±0.50 D Simulcross targets are substituted for the ±0.25 D ones.

In use, the Simulcross target system is complemented by the stepping capability of the SR-IV optics. Red and blue stepping buttons are provided immediately above the operator's control knob. Pushing the left-hand (red) button has the same effect as turning the control knob one step in the counter-clockwise direction. Pushing the right-hand (blue) button has the same effect as turning the control knob one step in the clockwise direction.

As previously noted, the two halves of the Simulcross targets are color-coded by red and blue peripheral bands. If the crosses on the red side are blacker and sharper, pushing the red stepping button makes the appropriate change, and if the crosses on the blue side are blacker and sharper, pushing the blue stepping button makes the appropriate change. There are no other conventions for the operator to remember; cross-cylinder refinement is thus made simple, fast, and dependable.

During cross-cylinder refinement of cylinder power, each push of either stepping button causes cylinder power to change 0.25 D, and simultaneously causes the sphere to change 0.23 D in the opposite direction. This automatic sphere compensation maintains the spherical equivalent constant, keeping the circle of least confusion of the conoid of Sturm on the retina, as required for accurate cross-cylinder refinement. It is actually cross-cylinder power that is changing, with both the SPHERE and the CYLINDER legends on the control panel being illuminated simultaneously.

The stepping buttons behave differently during cross-cylinder refinement of cylinder axis. In order to provide relatively uniform change in focus for the patient with each axis step, large axis steps are taken when small cylinders are present, and small axis steps are taken when large cylinders are present. If the cylinder power (C) in the SR-IV can be assumed to be correct, a quarter diopter worth of cross cylinder focus change will occur with an axis change of $\sin^{-1} (\frac{1}{8C})$ degrees. Relatively uniform focus changes are thus provided according to the axis step sizes given in Table 1. Uniform sensitivity of cross-cylinder power and axis testing is thus achieved regardless of the power of cylinder present. This built-in variable stepping feature is a powerful advantage unique to the SR-IV [6].

Table 1 Axis step sizes during cross-cylinder refinement

SR-IV cylinder value present	Axis step sizes
0.00 to less than 1.00 D	10°
1.00 to less than 2.00 D	5°
2.00 to less than 3.00 D	3°
3.00 to less than 5.00 D	2°
5.00 to 8.00 D	1°

The Simulcross system is responsible for the most important advantage of the SR-IV--accuracy. The SR-IV is designed to be more accurate than the phoropter. It has to be, for it must produce end-point refractions for operators who know little or nothing about optics.

Summary

Edward Jackson could not have foretold that patients would be refracted by looking into boxes in the 1980's. If he could look inside, however, he would be comforted to find the same lens which he found in 1887 was far more useful, and far more used, than any other single lens in his trial set.

References

1 S.J. Beach: Am. J. Ophthalmol. 11, 209 (1928)
2 W.J. Biessels: J. Am. Optom. Assoc. 38, 473 (1967)
3 D.L. Guyton: Am. J. Optom. Physiol. Opt. In press, 59 (1982)
4 D.L. Guyton: Ophthalmic test apparatus having magnification compensation. U.S. Patent pending.
5 P.R. Haynes: Am. J. Optom. Arch. Am. Acad. Optom. 35, 637 (1958)
6 M.H. Waltuck and D.L. Guyton: Control means for optical instrumentation. U.S. Patent pending.

Enhancing the Efficiency of Cross-Cylinder Astigmatism Testing

John H. Carter

The New England College of Optometry, Boston, MA 02115, USA

Introduction

The crossed cylinder was introduced by Stokes in 1849 and popularized three quarters of a century later by EDWARD JACKSON [1]. The crossed cylinder is most often used to refine axis and power of a presumptive cylinder determined by some method such as retinoscopy, and the testing involves the use of forced choice. Sensitivity is enhanced when differences in blur for paired stimuli are maximized for a given amount of error. Fogging is avoided since small amounts of accommodation are inconsequential and because the presence of any baseline blur diminishes contrast between paired presentations. Intentional overcorrection of the presumptive cylinder should be avoided [2,3].

The purposes of this paper are to: (1) examine factors that influence test sensitivity, (2) develop internally consistent tolerances for axis precision, and (3) suggest modifications of standard technique intended to enhance efficiency.

Basic Optical Considerations

When a cylinder is placed before an astigmatic eye, a spherical focus results only if: (1) the cylinder axis is aligned with a principal meridian of the eye, and (2) cylinder power is of correct sign and proper amount to fully collapse Sturm's interval. Should the axis of a presumptive cylinder of exact or inexact power be malaligned relative to the eye's principal meridian, the principle of obliquely crossed cylinders generates a residual astigmatic error whose power and axis reflect the vector sum of the astigmatic error of the eye and the ophthalmic cylinder before the eye.

Introduction of a crossed cylinder before an eye and trial cylinder results in a three-element combination whose spherical equivalent is unaltered by the crossed cylinder. The astigmatic residue may vary depending upon crossed-cylinder orientation. While astigmatic residuals may differ with respect to axis and power, it is power that largely determines any resulting acuity difference.

GARTNER [4] described a simple method to combine the effects of two or more cylindrical elements. Common vectors may be employed if axis values are doubled, thereby converting the 180° axis scale to a full circle. In Fig. 1, vector e represents the eye's astigmatic power error, and t is a trial cylinder of exact power, malpositioned by $\emptyset$. The double angle, $2\emptyset$, is shown in the figure. v_1 and v_2 are vectors corresponding to the first and second positions of a crossed cylinder. Since the optical axis of the crossed cylinder in each case forms an angle of 45° with the axis of the presumptive cylinder, the vectors must intersect at twice that angle (90°).

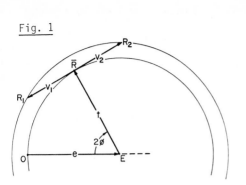

Fig. 1

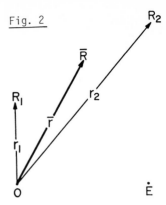

Fig. 2

Figure 1 Shows the effect of combining an ocular error cylinder e with a trial cylinder t and a crossed cylinder v. The vector lengths for e and t are the same, being represented in the figure by the directed lengths OE and E$\bar{R}$, respectively. v_1 and v_2 correspond to the initial and flipped vectors for the crossed cylinder. Their respective termini at R_1 and R_2 lie at equal distances from $\bar{R}$ but at differing distances from the zero error point, O. The angle constructed between each component pair is twice its corresponding physical value to permit the use of standard vector notation and analysis.

Figure 2 Reconstructs certain points of Fig. 1 in relationship to O, the zero-error point. Note that $\bar{r}$ defines the residual error for the trial-lens corrected eye when the axis error is Ø. r_1 and r_2 are the residual astigmatic errors for the three-element (eye, trial-lens, crossed-cylinder) combination for the first and second positions of the crossed cylinder, respectively. $(r_2 - r_1)$ represents the corresponding arithmetic difference in resultant astigmatic error.

Figure 2 should be referenced to Fig. 1, r is a vector defining the distance from the zero error point to the terminus of the trial cylinder vector. It reflects the residual astigmatic error for the trial-cylinder corrected eye. Vectors r_1 and r_2 apply with the crossed cylinder in its two respective principal orientations. For some value of Ø, r_1 reaches a minimum value. I have termed this value of Ø the <u>critical angle</u>. When the error in axis placement is smaller than the critical angle, points R_1 and R_2 lie on opposing sides of the zero-error point O as illustrated in Fig. 3. Then, $(OR_2 - OR_1)$ assumes a value only slightly less than twice the value of O$\bar{R}$. This enhanced blur difference lends sensitivity to crossed-cylinder axis testing.

I have defined dioptric contrast by the formula

$$C_\% = 100^* \ (r_2 - r_1)/r_2 \ . \tag{1}$$

Figure 4 shows how dioptric contrast varies with axis error for the first and second crossed-cylinder positions. It is based upon a strong (2.50 D) error and crossed-cylinder powers of ±0.12 and ±0.37 D. Note that dioptric contrast increases with axis error up to the critical angle, whereafter it declines.

Figure 5 presents additional data for elements of Fig. 4. The curve marked with a cross shows the astigmatic residual $\bar{r}$ for a trial-cylinder corrected eye. The curves drawn with filled and open circles show sequential dioptric

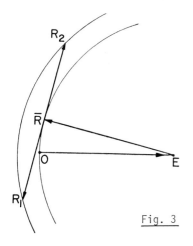

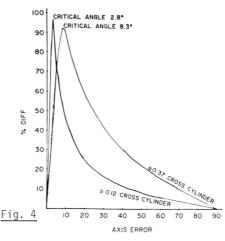

Fig. 3   Fig. 4

Figure 3 Dioptric contrast has been defined by formula (1). Its value as-
cends rapidly with increasing axis error when R_1 and R_2 lie on opposing sides
of the zero-error point, the condition demonstrated here. When axis error
becomes large so R_1 and R_2 lie on the same side of the zero error point,
dioptric contrast begins to fall.

Figure 4 Shows how dioptric contrast varies with axis error for two and one-
half diopters of astigmatism tested with a 2.50 D trial cylinder and crossed
cylinders of ±0.12 and ±0.37 D, respectively.

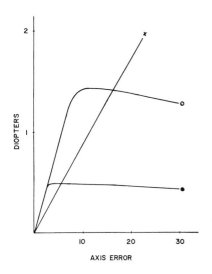

Figure 5 Shows (plotted with X) how
residual astigmatic error $\bar{r}$ varies with
axis error for 2.5 D of astigmatism and
an equal-valued trial cylinder. The same
figure shows how arithmetic dioptric
difference (r_2-r_1) for the residual cylinder varies with axis error when a
crossed cylinder is used to refine
axis. Curves plotted with open and
filled circles reflect sequential dioptric differences corresponding to
±0.37 and ±0.12 D crossed cylinders,
respectively

difference (r_2-r_1) values as a function of axis error for ±0.12 and ±0.37 D
crossed-cylinder powers, respectively. Note that (r_2-r_1) grows almost twice
as fast as $\bar{r}$ up to the critical angle. Thereafter, growth of (r_2-r_1) levels
off and eventually reverses. Figure 5 also shows that for a 2.5 D cylindrical error, (r_2-r_1) is much smaller for the ±0.12 compared to the ±0.37 D
crossed cylinder. Even though a low power cross cylinder may yield favor-

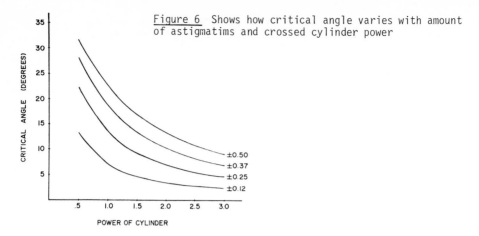

Figure 6 Shows how critical angle varies with amount of astigmatims and crossed cylinder power

able dioptric contrast for small axis errors, the absolute values of r_2 and r_1 and the difference between these then may all be small compared to the eye's dioptric depth of focus. Such small absolute differences may fail to be detected even though dioptric contrast is high.

When error in axis placement is equal to or less than the critical angle, the dioptric contrast is large and the absolute difference between power errors for the two crossed-cylinder positions is nearly double the power error that results from the displaced trial cylinder alone. Attempts to refine a presumptive axis that forms a large angle with the axis of the ocular astigmatism results in low test sensitivity and thus a potential for failing to proceed directly toward the required axis location. Figure 6 shows how critical angle varies as a function of astigmatic error and crossed-cylinder power.

Establishing Axis Tolerance

Ocular depth of focus, the use of standard dioptric intervals, and other factors necessitate tolerance. Important factors that affect tolerance values include: (1) ocular factors (e.g., clarity of the media and integrity of the retina), (2) general patient factors (e.g., intelligence, observational ability, willingness to accept blur), and (3) situation-specific elements (e.g., occupational, avocational, and other visual requirements). While a residual cylinder can derive in part from an error in presumptive cylinder power and in part from axis error, it often is convenient for the clinician to view axis and power errors independently. It is relatively easy to set tolerance levels for power error but far more difficult to establish internally consistent tolerances for axis error. Axis error is important principally by virtue of the power error it produces. We can define the relationship between tolerance to power error and corresponding axis error tolerance by the formula:

$$T_a = Sin^{-1} (T_p/2*C)) ,\qquad\qquad (2)$$

where T_a and T_p are cylinder axis and power tolerance, respectively, and C is the amount of astigmatism.

Given an average patient with good corrected acuity many practitioners will attempt to determine cylinder power to within ±0.25 D. Table 1 is based

92

Table 1 Allowable axis error as a function of amount of astigmatism. Figures are based upon the assumption that ±0.25 D represent acceptable limits for astigmatic power error.

A. Cylinder Power	B. Allowable Axis Error	C. Clinical Approximation	D. Exact Power Error Resulting From Approximation
0.50 D	±14.48°	±15°	.259 D
1.00 D	± 7.18°	± 7.5°	.261 D
1.50 D	± 4.78°	± 5°	.261 D
2.00 D	± 3.58°	± 4°	.279 D
2.50 D	± 2.86°	± 3°	.262 D
3.00 D	± 2.39°	± 2.5°	.262 D
3.50 D	± 2.05°	± 2°	.244 D
4.00 D	± 1.79°	± 2°	.279 D

Table 2 Various data, arranged by cylinder power, for the first, second, third, and fourth crossed-cylinder presentation.

Cylinder Power	Data Item	Flip Axis Position Number			
		1	2	3	4
0.50	(A)	0	15°	45°	–
	(B)	p = ~.5	p = .51	p = .88	–
1.00	(A)	0	7.5°	22.5°	37.5°
	(B)	p = ~.5	p = .52	p = .92	p = .97
1.50	(A)	0	5°	15°	25°
	(B)	p = ~.5	p = .52	p = .92	p = .98
3.00	(A)	0	2.5°	7.5°	12.5°
	(B)	p = ~.5	p = .52	p = .92	p = .98

upon this value. A more liberal axis tolerance is indicated whenever power error tolerance would be increased, as in the case of the unresponsive patient or one whose visual acuity is limited by nondioptric factors. And, on rare occasions lower tolerances may be indicated. In Table 2, Columns A and B present values for amounts of astigmatism and correlated axis tolerance, respectively. Column C rounds numbers of Column B to clinically convenient values. Column D shows actual dioptric power errors resulting from the use of rounded values.

Strategies for Crossed-Cylinder Tests

In terms of search strategy, axis refinement can be reduced to selecting the best from among n axis numbers, where n depends upon several factors including cylinder strength. It is convenient to define families of axis numbers so that each is designated by its most central value; now the particular values, 180° and 90°, always fall within the number set. If tolerance for induced power error is taken as one-fourth diopter, only six axis numbers are required for a one-half diopter cylinder. Twelve are needed for a one dioptic cylinder and eighteen are required for a one and one-half diopter cylinder.

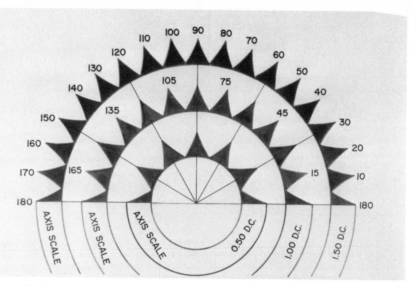

<u>Figure 7</u> Shows how axes can conveniently be grouped if allowable power error for the residual cylinder is taken as ±0.25 D. This figure has been constructed for 0.50, 1.00, and 1.50 D of astigmatism.

Thirty-six axis numbers must be employed to define the orientation of a three diopter cylinder. Figure 7 is a graphical representation, based upon tolerances shown in Table 1, which assigns convenient axis numbers to 0.50, 1.00, and 1.50 D cylinders. Values for a 3.00 D cylinder are not shown but fall at 5° intervals that correspond to the standard refractor's axis scale. For intermediate cylinder powers, it probably is best to use values assigned to the next most powerful cylinder. This constrains residual power error within original tolerances.

The average number of trials needed to refine axis or power can be minimized through the use of a procedure that renders equally likely the preference for first and second crossed-cylinder orientations. To establish conditions expected to lead to such equiprobability, however, seems appropriate only for the first two pairs of crossed cylinder trials. This is because, in the vast majority of cases, the third pair of trials can be expected to yield an acceptable end point.

In axis refinement, the axis of the trial cylinder is adjusted to a standard value. Refer again to Fig. 7. Then, the first pair of crossed-cylinder presentations is made in standard fashion, the axis of the trial cylinder and the crossed-cylinder flip axis coinciding with the presumed axis for the ocular astigmatism. If equality is not realized on this initial test, the working cylinder axis and crossed cylinder flip axis are displaced in the appropriate direction. Distance of displacement depends upon presumed cylinder power with the displacement angle equal to half the size of the standard axis interval for a cylinder of corresponding power. Thus, the displacement angle should be 15° for a 0.50 cylinder and 7.5°, 5°, and 2.5° for 1.00, 1.50, and 3.00 D cylinders, respectively. The axis test is then repeated at this new location.

94

If the outcome of this second test is directed toward the original presumptive axis, cylinder-axis refinement is complete. And, of course, equality indicates the true axis to be sited at the test axis. Otherwise, a third pair of crossed-cylinder presentations is indicated. This time, the working cylinder and crossed-cylinder flip axes are displaced by an additional amount equal to the width of the axis interval. On this third trial, there is considerable likelihood that the patient's response will be directed toward the original presumptive axis, in which case the recorded axis is that which designates the second interval. Reported equality places the true axis at the test axis location but a response directed away from the original presumptive axis must be evaluated in the context of cylinder strength. Recall that test sensitivity can be extremely low when the error angle is large. A large initial error angle can result in failure of a test sequence to lead systematically toward the proper axis location and it sometimes is best for the practitioner to re-evaluate his starting point rather than to continue with the refinement process. For example, the third test with a half-diopter cylinder will have been conducted 45° from the original axis so a response directed toward an even greater deviation suggests a need to re-evaluate the original estimate of cylinder presence and location.

For a cylinder believed to be one diopter or more in power, a fourth test is indicated. It consists of iteration of the procedure described above at a test eccentricity increased by the size of the standard axis interval. Failure to confirm that the true axis lies within the third interval probably suggests a need to re-evaluate retinoscopic and/or other preliminary data.

Table 2 shows various data, arranged by cylinder power for the first, second, third, and fourth crossed-cylinder presentations. (A) indicates test eccentricity by trial number for each cylinder power. (B) indicates the corresponding theoretical probability of a response directed toward the original presumptive axis. The p values are based upon an assumed standard deviation of ±0.37 D for recognition of axis-induced power error. Appreciable deviation from equiprobability does not occur for the second pair of presentations until the assumed standard deviation falls outside the limits of ±0.25 and ±0.62 D. These values yield balances approximating 70:30 and 30:70, respectively.

Power refinement follows axis refinement and is more simple and straightforward. Comparisons are made with the flip axis 45° from the trial cylinder axis. The power error difference for the two positions of the crossed cylinder is expected to be twice the residual power error present without the crossed cylinder if this is no greater than the astigmatic component of the crossed cylinder and if the axis is correct. If equality of clarity is obtained on the first set of trials, the tentative cylinder can be assumed correct. If the first set of trials indicates the possible need for a stronger or weaker cylinder, the trial cylinder is strengthened or weakened one-fourth diopter. If we again assume a ±0.37 D standard deviation for recognition of power error, this 0.25 D adjustment in presumptive cylinder power should lead to a crossed-cylinder test outcome that is essentially equiprobable for clarity-based preference for the first versus the second presentation. If the patient rejects this alteration in power, the original estimate of cylinder power is confirmed. If equality exists for the two crossed-cylinder positions, the trial cylinder is of correct power. Should the outcome indicate still greater deviation in power from the original cylinder, a third set of trials may be conducted following an additional one-fourth diopter change in trial cylinder power. This time, the probability of a response directed toward the original power value is about 0.9.

Summary and Conclusions

Certain important attributes of crossed-cylinder tests for axis and power are well known. The spherical equivalent of a crossed cylinder is zero, so flipping a crossed cylinder does not alter the equivalent sphere for any combination of lenses before the eye. The crossed cylinder lends itself to use of the method of forced choice, a technique valued by refractionists since it is economic in terms of time and since it does not impose excessive demands upon a patient of modest observational ability.

A less generally recognized advantage to the use of the crossed cylinder is its high potential sensitivity when the axis error is small. When axis position error is small, the absolute difference between residual power errors for a set of crossed-cylinder presentations is almost twice the residual power for the eye and trial-lens combination alone. However, attempts to refine a presumptive axis that forms a large angle with the axis of the ocular astigmatism results in low test sensitivity and therefore gives rise to the possibility that testing may fail to lead systematically toward the proper axis location. The axis error value at which test sensitivity begins to diminish with further increase in axis error has been termed the critical angle. The value of the critical angle varies directly with crossed-cylinder strength and inversely with amount of astigmatism.

Efficient use of the crossed cylinder can be achieved by establishing tolerances for axis error and by attempting only to reduce residual error to within these tolerances. This approach can be facilitated by pre-establishing an axis scale. The scale is determined from the estimated power of the cylinder and the established power tolerance. It is convenient to use a scale which includes the values 90° and 180°. For the average patient with good corrected acuity, a tolerance for axis-dependent residual power error of ±0.25 D probably represents a good choice. Then, only six numbers are required to accommodate the axis designations for half-diopter cylinders. Twelve and eighteen numbers are required for one diopter and one and one-half diopter cylinders, respectively. Thirty-six numbers are needed in the case of a three-diopter cylinder.

With the use of a preselected axis scale, determination of the "correct" axis reduces to the selection of one from among n values. Efficiency in testing can be enhanced if expectations as to preferred response can be made approximately equal for the first and second crossed-cylinder orientations. While this procedure is practical only for the first two pairs of comparisons, final determination usually is complete within three sets of trials.

Actually, the residual power error resulting from use of the described method should be somewhat less than the nominal tolerance because failure to recognize a difference in clarity is taken to indicate that the axis is sited at the test axis location.

Power refinement using the crossed cylinder is more simple and straightforward than axis refinement. When using the crossed cylinder to refine power, the difference between the residual power errors for the first and second crossed-cylinder positions can be expected to be double the residual error for the eye and trial cylinder alone if this error does not exceed the astigmatic component of the crossed cylinder and if the axis is correct. No modification of the standard method for power refinement is suggested.

Errors in cylinder axis and power are not entirely independent as clinicians often assume them to be, but coexisting errors giving rise to a single residual error cylinder as their vector sum. Such interactive effects are interesting but beyond the scope of this paper.

References

1 E. Jackson: Am. J. Ophthal. 13, 321 (1930)
2 J.H. Carter: Optometric Monthly 72, 15 (1981)
3 Irvin M. Borish: *Clinical Refraction*, 3rd Ed. (Professional Press, Chicago, 1970)
4 W.F. Gartner: Am. J. Optom. and Arch. Am. Acad. Optom. 42, 459 (1965)

Parameters of the Change in Clinical Refractive Error in Young Myopes

David A. Goss

College of Optometry, Northeastern State University, Talequah, OK 75564, USA

Introduction

A common clinical finding in youngsters with myopia is an increase in the amount of the myopia over a period of months or years, as manifested by an increase in the minus sphere spectacle lens required to correct to best visual acuity. Conventional wisdom in the ophthalmic professions holds that this myopia progression continues until the middle or late teens. This has been confirmed in general terms in papers by BUCKLERS [1] and HOFSTETTER [2] but it has not been studied previously on a quantitative basis.

Various papers have shown sex differences in different aspects of myopia development. Work by YOUNG et al. [3] and HIRSCH [4] indicates that myopia tends to appear earlier in females. HIRSCH [5] found that in over 5,000 clinic records for patients aged 18 to 50 the prevalence of myopia greater than 6 D in females was twice that in males. A possible sex difference in the rate of progression of myopia is supported by a study by ROSENBERG and GOLDSCHMIDT [6]. They reported that of 108 boys and 144 girls seen in a private practice, about 5% of the boys and 10% of the girls increase in myopia more than 2 D in the two years after the onset of myopia.

Numerous reports have also appeared in the literature suggesting that the progression of myopia can be favorably altered by various treatment schemes [7], the most popular of which has been bifocal lenses. However, reports presenting data concerning the use of bifocals have not shown consistent success [7,8].

The purpose of this paper is to examine 1) the ages of cessation of myopia progression, 2) the rates of myopia progression, 3) sex differences in these two parameters, and 4) the rate of progression as a function of the two types of spectacle lens prescription.

Subjects and Data

Clinical records of 299 patients were selected from three optometry practices in Iowa, Illinois, and Indiana, based on the following criteria: 1) at least four refractive examinations between the ages of 6 and 24 years, 2) myopia of at least 0.5 D sometime during the course of the record, 3) astigmatism never manifested in excess of 2.5 D, 4) no strabismus or amblyopia, 5) no contact lens wear prior to the last examination recorded for this study, 6) no ocular pathology, and 7) no systemic pathology which might affect ocular findings. Among the data recorded for each examination was the subjective refraction, that is, the cylindrical correction combined with the least minus spherical lens power required for best visual acuity. Further information concerning the subjects and data is given in another paper [9].

Methods

After plotting linear coordinate graphs of amount of myopia vs. age for these patients it became obvious that the majority of individuals showed a plot which could be interpreted as a linear increase in the degree of myopia into the middle or late teens, followed by a relatively flat, horizontal line segment on the graph. To illustrate this, Fig. 1 shows representative plots from the male sample, and Fig. 2 shows plots selected from the female sample. To derive cessation ages and rates of progression of myopia a linear model of two lines, one with a more negative rate of change and the other with a zero or near zero slope was used. The validity of this linear model was verified when a statistical test for linear regression was applied to the observations during the years when the myopia was increasing, and the relationship between refractive error and age was given by a straight line in the majority of cases [10].

Cessation age: For an index of the cessation age or the theoretical termination point of the progression of myopia, four related methods were used. These four methods and the results obtained with them will be described in another paper [9]. For the sake of brevity, only two of those methods will be discussed here.

Method 2 involved fitting a straight line through the ascending portion of the graph of the amount of myopic refractive error in the principal meridian nearest horizontal in the right eye vs. age, by linear regression analysis. At least three points before 15 years of age were required for inclusion in this analysis. The age at which this regression line intersected a zero slope line through the mean refractive error for points after 17 years of age was taken as the cessation age. The results with Method 1, omitted here, were highly correlated with those of Method 2 [9].

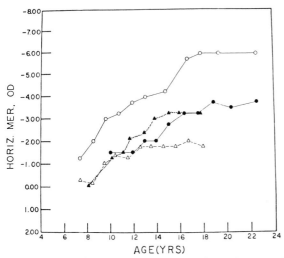

Fig. 1 Graphs of male subjects selected as typical and representative of the sample. Each set of common symbols represents one individual. Age in years is on the X-axis, and diopters of refractive error in the principal meridian nearest horizontal in the right eye is on the Y-axis.

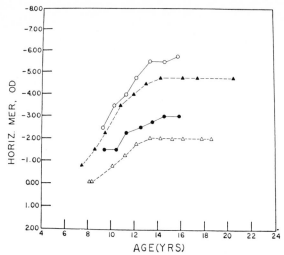

Fig. 2 Graphs of female subjects selected as typical and representative of the sample. Each set of common symbols represents one individual. Age in years is on the X-axis, and diopters of refractive error in the principal meridian nearest horizontal in the right eye is on the Y-axis.

Method 4 used the mean of the spherical equivalent refractive errors of the two eyes. For the determination of cessation age, this method involved fitting the refractive error vs. age data by a pair of regression lines without age restrictions in the construction of the lines and without restricting the second line to zero slope [9]. The age at which the regression lines met was taken as the cessation age. In general, for inclusion in the analysis by Method 4, a record had to have contained points covering an age period of at least 11 to 16 years of age. The results with Method 3, omitted here, were highly correlated with those of Method 4.

Data Analysis

Rates of increase of myopia were given by the slopes obtained from the linear regression analysis of the first line segment. A positive slope in graphs such as Figs. 1 and 2 corresponds to a negative rate of change in diopters of refractive error per year; that is, it corresponds to an increase in myopia. Rates were compiled by two methods, which we shall call Methods A and B.

Method A included rates obtained 1) from the slopes of the first regression line from Method 2 used for cessation ages, and 2) from the slopes of linear regressions of refractive error in the principal meridian nearest horizontal in the right eye vs. age, for subjects whose last observation was at 16 years of age or earlier. Method B included rates obtained 1) from the slopes of the first regression line from Method 4 used for cessation ages, and 2) from the slopes of linear regression of binocular mean spherical equivalent refractive error vs. age for all other subjects.

Results

The mean cessation ages by Method 2 were 16.66 years for males and 15.21 years for females. Statistical significance of the differences of these means was established at the 0.0001 level. Mean cessation ages by Method 4 were 15.53

years for males and 14.57 years for females. These means were significantly different at the 0.02 level. For both males and females, there was considerable variability in cessation age. The standard deviations of cessation age were in the neighborhood of 2 years for both sexes and both methods (see Table 1).

The mean rates for refractive error change, in D /yr, from Method A were -0.45 for males and -0.50 for females. The difference in these means was not statistically significant at the 0.05 level. The Mann-Whitney U test of rank sums [11] also did not yield a significant difference. The mean rates from Method B were -0.38 D./yr for males, and -0.46 D /yr for females (see Table 2). This difference in means was significant at the 0.02 level. The Mann-Whitney U test showed a significant difference in the groups at the 0.02 level, as well. The frequency distributions of rates in Table 3 suggests that there are more females at higher rates of myopia increase. For example, from Method A, 1 (1%) of 101 males had a rate of myopia increase greater than -0.92 D /yr, whereas 6 (6.5%) of 93 females had a rate more negative than -0.93. Further statistical analysis is being planned to investigate the possible difference in rates between males and females.

Information was also available in this data pool concerning the use of bifocal lenses for young myopes, since two of the three offices from which the data were taken used bifocals with some regularity for nearsighted youngsters. In one office, the bifocal used was typically a Kryptok with a +0.75 or +1.00 add placed with the top of the segment 1 mm below the geometrical center of the frame. In the other office a flat-top segment with a +1.00 or +1.25 add placed 1 mm "below" the bottom edge of the pupil was the bifocal most typi-

Table 1 Mean cessation ages, standard deviations, and levels of significance of difference of mean cessation ages for males and females.

METHOD	SEX	N	MEAN	S.D.	SIGNIFICANCE
2	M	66	16.66	2.10	0.0001
	F	57	15.21	1.74	
4	M	59	15.53	1.93	0.02
	F	45	14.57	2.18	

Table 2 Mean rates of refractive error change (in D/yr) and standard deviations for males and females.

METHOD	SEX	N	MEAN	S.D.
A	M	101	-0.45	0.22
	F	93	-0.50	0.25
B	M	159	-0.38	0.25
	F	131	-0.46	0.29

Table 3 Frequency distributions of rates of refractive error change (D/yr) for males and females. The full range of rates in each case was divided into five equally sized intervals.

RANGE FOR MALES	NO. OF INDIVIDUALS	RANGE FOR FEMALES	NO. OF INDIVIDUALS
Method A		Method A	
-1.15 to -0.92	1	-1.52 to -1.22	1
-0.92 to -0.69	15	-1.22 to -0.93	15
-0.69 to -0.45	30	-0.93 to -0.63	18
-0.45 to -0.22	40	-0.63 to -0.34	42
-0.22 to +0.01	15	-0.34 to -0.04	27
Method B		Method B	
-1.30 to -1.03	3	-1.42 to -1.12	3
-1.03 to -0.77	7	-1.12 to -0.82	12
-0.77 to -0.50	40	-0.82 to -0.52	34
-0.50 to -0.24	58	-0.52 to -0.22	51
-0.24 to +0.03	51	-0.22 to +0.08	31

Table 4 Mean rates (D/yr) and standard deviations for single vision lens wearers (SV) and bifocal lens wearers (BF) for males (M) and females (F) considered both separately and together.

METHOD	LENS	SEX	N	MEAN	S.D.
A	BF	M	16	-0.44	0.22
	SV	M	21	-0.50	0.27
A	BF	F	11	-0.48	0.30
	SV	F	17	-0.44	0.18
A	BF	M & F	27	-0.45	0.25
	SV	M & F	38	-0.48	0.23
B	BF	M	21	-0.45	0.25
	SV	M	38	-0.35	0.26
B	BF	F	18	-0.44	0.36
	SV	F	22	-0.36	0.20
B	BF	M & F	39	-0.44	0.30
	SV	M & F	60	-0.36	0.24

Table 5 Mean rates (D/yr) and standard deviations for single vision lens wearers (SV) and bifocal lens wearers (BF) with different levels of ACA ratio.

METHOD	LENS	ACA	N	MEAN	S.D.
A	BF	< 6	13	-0.50	0.22
	BF	$\geq$ 6	14	-0.42	0.28
	SV	< 6	24	-0.46	0.25
	SV	$\geq$ 6	13	-0.50	0.20
B	BF	< 6	17	-0.52	0.30
	BF	$\geq$ 6	22	-0.38	0.29
	SV	< 6	38	-0.36	0.25
	SV	$\geq$ 6	22	-0.35	0.22

cally employed. Using data only for those individuals who wore just one of the lens types from the first to last observations used in the determination of the rate, the mean rates for single vision lens wearers and bifocal lens wearers were not significantly different at the 0.05 level for either the sexes considered together or separately by either Method A or B (see Table 4). The Mann-Whitney U test did not show significant differences between the single vision and bifocal groups.

Rates with single vision lenses and bifocals as a function of ACA ratio were also investigated. Each lens group was divided into two ACA ratio groups. This division was arbitrarily made at the 6:1 ACA level. Statistically significant differences were not found between the means of any of these groups (see Table 5). The Mann-Whitney U test yielded a significant difference in one case here. With the data from Method A, bifocal wearers with ACA ratios less than six were shifted toward higher rates as compared to single vision patients with ACA ratios less than six (significant at 0.01 level).

An incidental analysis was also made of the correlation of rate with two parameters of accommodation and convergence. The correlation coefficient of rate vs. ACA ratio was 0.03. The correlation coefficient of rate vs. the Positive Relative Accommodation test at 40 cm obtained at the first examination was 0.11. This suggests that there is no obvious relationship between these two factors and the rate of increase of myopia.

Discussion

The cessation age results may be of aid to the ophthalmic practitioner in answering inquiries from patients and parents about when myopia will stop increasing. Using the results from Method 2, one could say that boys are expected to stop at about 16 2/3 years of age, and girls at about 15 1/4 years, although there is a great deal of variability, with differences of 2 or 3 years in either direction from these ages not being uncommon.

The results of this study support the idea that there are sex differences in myopia progression. There is the theoretical possibility that a more rapid and earlier ending progression of myopia in females as compared to males may be correlated with the faster and earlier ending general body growth of females. This still would not tell us whether the etiology of myopia is predominantly hereditary or environmental. That is, the genetically determined changes in refractive error might simply end, or the environmental determinants of refraction may no longer be effective, after growth ceases.

The results presented here do not support the theory that bifocals can slow the progression of myopia. ROBERTS and BANFORD [12] suggested that bifocals may be more effective in reducing the rate of myopia increase for certain individuals, especially females and persons with high ACA ratios. Although sample sizes were small for these subgroups in the present study, the results do not support such a hypothesis. ROBERTS and BANFORD [12] also indicated that youngsters with low ACA ratios may change faster with bifocals than with single vision lenses. The results here suggest that this bears further investigation.

Summary

A common clinical finding in youngsters with myopia is an increase in the myopia until the middle or late teens. This is manifested as an increase in the minus sphere spectacle lens needed to correct for best visual acuity. To study the parameters of this phenomenon, longitudinal data for 299 young myopes, examined at least four times between the ages of 6 and 24 years, were obtained from three optometry practices. A model of two straight lines, one with a higher negative slope (in D/yr) and another with a lower negative or zero slope, was used to describe the change in refractive error with age. Using this linear model an index of the age at which myopia stops increasing was determined. The mean myopia progression cessation age for females is significantly lower than that for males. The linear model was also employed to determine rates of refractive error change by regression analysis. The mean rate was more negative for females than for males. Mean rates for single vision lens wearers and bifocal lens wearers were not statistically significantly different. Etiological and clinical implications are discussed.

Acknowledgments

The author would like to thank the following persons for assistance and/or suggestions at various stages of this work: Drs. H.W. Hofstetter, R.L. Winkler, D. Cox, V. Metzger, P. Randolph, A. Lande, D. Crouch, G. Fulk, P. Erickson, and Mr. J. Carter. Partial support was provided by the Indiana University Rudy Professorship Research Fund, and an institutional grant from Northeastern State University.

References

1   M. Bucklers: Brit. J. Ophthalmol. 37, 587 (1953)
2   H.W. Hofstetter: Am. J. Optom. Arch. Am. Acad. Optom. 31, 161 (1954)
3   F.A. Young, R.J. Beattie, F.J. Newby, and M.T. Swindal: Am. J. Optom. Arch. Am. Acad. Optom. 31, 111 (1954) and 31, 192 (1954)
4   M.J. Hirsch: The refraction of children; in *Vision of Children*, ed. by M.J. Hirsch and R.E. Wick (Chilton, Philadelphia 1963)
5   M.J. Hirsch: Am. J. Optom. Arch. Am. Acad. Optom. 30, 135 (1953)
6   T. Rosenberg and E. Goldschmidt: Doc. Ophthalmol. Proc. Ser. 28, 33 (1981)
7   D.A. Goss: Am. J. Optom. Physiol. Opt. 59, 828 (1982)
8   T. Grosvenor: Optom. Monthly 71, 620 (1980)
9   D.A. Goss and R.L. Winkler, Progression of myopia in youth: age of cessation, submitted manuscript.

10 D.A. Goss and R.L. Winkler: Linearity of myopia progression in childhood. Working paper, to be submitted.
11 R.G.D. Steel and J.H. Torrie: *Principles and Procedures of Statistics* (McGraw-Hill, New York 1960)
12 W.L. Roberts and R.D. Banford: Optom. Weekly 58, 25; 58, 21; 58, 23; 58, 27; 58, 19, 26 (1967)

Deformed Aspheric Ophthalmic Lenses

Milton Katz

SUNY State College of Optometry, New York, NY 10010, USA

1. Introduction

A previous study [1] indicated the feasibility of minimizing astigmatic error, power error, and oblique astigmatic error in -14 and -20 diopter ophthalmic lenses. Attempts to minimize these errors in a +14 diopter lens were unsuccessful. The aims of the present study are to determine whether any positive lenses between +6 and +14 diopters can be asperized to minimize astigmatic error, power error and distortion, and the general form of negative lenses that are satisfactorily asperized. Lenses of the following dioptric power are included in this study: -20, -14, -10, -8, +6, +8, +10, and +14 D.

Spherically surfaced minus lenses and low powered plus lenses can be bent to correct astigmatic error or optimally balance astigmatic and power errors. However, distortion cannot be corrected alone or in conjunction with these other errors if spherical surfaces are employed [2-7]. Only one degree of freedom is available to optimize the lens since ophthalmic lenses are singlets worn at a fixed distance from a rear stop presumed to be at the center of rotation of the mobile eye. In effect, one surface curvature can be varied. The other surface curvature of the lens must satisfy the required back vertex power F.

The interval between the tangential (t') and sagittal (s') foci on the chief ray shown in Fig. 1 is oblique astigmatism. The distances from the vertex sphere, centered on the center of rotation R, to these foci are the tangential focal length ($f'_t = vt'$) and sagittal focal length ($f'_s = vs'$). The reciprocals of these distances in meters are the tangential (F'_t) and sagittal (F'_s) powers in diopters. Oblique astigmatic error is

$$OAE = F'_t - F'_s .$$

Third-order correction of OAE can be obtained with a shallow and a steep lens bending. These bendings cause the sagittal and tangential foci to coincide with the Petzval surface, and produce a curved sharp image surface. Such point focal designs will generally exhibit power errors because the ideal image surface, the far point sphere, does not ordinarily coincide with the Petzval surface. Mean oblique power error is defined as:

$$MOE = \frac{F'_t + F'_s}{2} - F .$$

Power error can be eliminated by causing the sagittal and tangential foci to straddle the far point sphere, but this will reintroduce an astigmatic error. Distortion is indicated in Fig. 1 as the difference in intersection heights

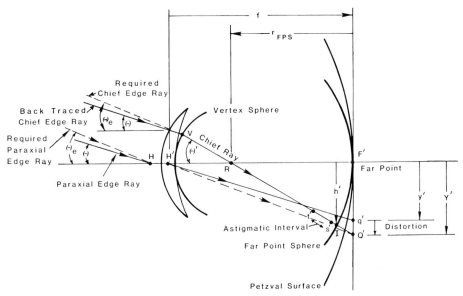

Fig.1. Optical geometry of a positive ophthalmic lens. The center of rotation of the eye is at R. This represents the location of the aperture stop for the lens. Rotation of the eye through the angle θ' dictates the slope of the chief ray in image space and the position of the image point Q'. The slope θ of the chief ray in object space is found by tracing the chief ray backward through the lens. The slope θ varies with the lens bending. The position of the par-axial image point q' is found by taking the paraxial edge ray with slope θ through the second principal point H'. There is no bending for spherically surfaced lenses that will produce an object space chief ray slope angle θ that will cause q' to coincide with Q'. The require chief ray angle θ$_e$ is given by the slope of ray H'Q' (broken line). The design goal is to find lens surfaces that will refract an incident ray with slope θ$_e$ through the center of the aperture stop (θ' = 30°) to the point Q' and also shift the s' and t' focuses to coincide at I on the far point sphere

of a chief ray and a paraxial ray at the transverse plane through the second focal point F'. Shallow point focal lens forms exhibit more distortion than the steep forms. Although distortion does not affect image sharpness, straight lines appear curved when viewed by the stationary eye through the lens periphery, and objects appear to move when the head is turned.

2. Point Focal Lens Forms

The unbroken line in Fig. 2 is a trigonometrically ray-traced version of the Tscherning ellipse for thick lenses made of ophthalmic crown glass (523589). The curve represents bendings of second surface power versus back vertex power for point focal lenses. The shallow point focal bendings are termed Ostwalt forms and the steep bendings are Wollaston forms. Point focal bend-ings exist for lenses between approximately +6.6 and -24.6 diopters. Spher-ical lenses that are stronger than +6.6 diopters cannot be made point focal, but can be bent for minimum astigmatic error as indicated by the broken line. The dot-dash line indicates bendings for minimum distortion lenses. These forms still exhibit appreciable distortion, as well as astigmatic and power errors. They are more steeply bent than the Wollaston point focal forms.

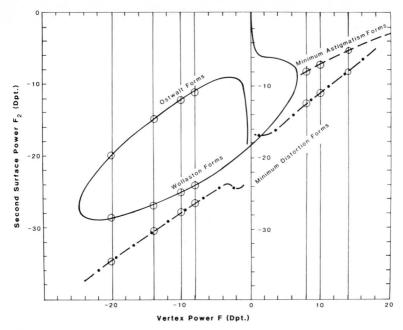

Fig.2. Trigometrically traced thick ophthalmic lenses corrected for astigmatic errors (point focal) are indicated by the unbroken line. A shallow (Ostwalt) form and a steep (Wollaston) form can be found for each vertex power between +6.6 and -24.6 diopters. Lenses of power greater than +6.6 diopters cannot be made point focal. The broken line indicates bendings at which these lenses exhibit minimal astigmatism. Distortion cannot be corrected for any lens power. However, the dot-dash line indicates bendings at which distortion is minimal. All the computations were made for spherically surfaced lenses made of ophthalmic crown glass and a vertex distance VR = 27 mm. Lens center thicknesses ranged from 0.5 to 9.0 mm for high minus to high plus lenses

3. Design Conditions

For reference purposes the astigmatism, power error and distortion of best form lenses were computed. Best form lenses are the Ostwalt and Wollaston point focal, and the minimum astigmatism and minimum distortion forms indicated by circles in Fig. 2 for the selected lens powers. The minimum distortion form is loosely termed a best form lens. Computations were made across the entire half field corresponding to 30° of eye rotation. The second surface of the lens was located 27 mm from the center of rotation of the eye which serves as a rear stop for the system. Lens center thicknesses ranged from 0.5 to 9.0 mm for high minus to high plus lenses of ophthalmic crown glass, n_d = 1.523.

4. Design Geometry

The basic optical geometry is illustrated in Fig. 1. A chief ray in image space with a slope θ', corresponding to the angle of eye rotation about the center of the aperture stop at R, intersects the flat paraxial image surface at Q'. This ray is traced backward through the lens to find the slope θ of the chief ray in object space. The slope θ in object space will depend upon the bending chosen for the lens. The position of the paraxial image point q' is determined by the intersection of a paraxial ray of slope θ drawn from the second principal point H' to the flat image surface. When θ is 30° the above rays are chief edge rays and paraxial edge rays.

Linear distortion is equal to Y'-y', or the distance between Q' and q'. Percent distortion is equal to $100(Y'-y')/y'$. Distortion will be eliminated if the slope of the chief ray in object space can be made equal to θ_e, where θ_e is the required slope of a paraxial ray through H' that intersects Q' as shown by the dashed line in Fig. 1. There is no bending of spherically surfaced lenses that will refract an incident object space chief edge ray with slope θ_e through the center of the stop with the required 30° slope. Consequently, all ophthalmic lenses exhibit distortion unless they are suitably aspherized.

Aspheric surfaces will eliminate astigmatic and power errors as well, if the sagittal and tangential foci can be made to coincide with I on the far point sphere. Similar conditions must be satisfied for intermediate field angles to correct these errors across the entire field of view.

5. Optimization

The first step in the optimization process was to select a lens bending (generally the minimum distortion form) for a given lens power and determine the intersection height Y' of the 30° image space chief edge ray. The slope of the chief edge ray in object space, required to correct distortion, was then obtained by finding:

$$\tan \theta_e - Y'/H'F' = Y'/f \text{ ,}$$

where f is the equivalent focal length. Since $\tan \theta' = Y'/RF'$, the ratio

$$\frac{\tan \theta'}{\tan \theta_e} = \frac{f}{RF'}$$

is a constant.

Distortion across the entire field of view will be corrected by satisfying this tangent condition for all chief zonal rays. Therefore, in addition to the 30° edge ray, five fractional field angles in image space were calculated. The slopes θ'_z of these chief zonal rays were found from: $\tan \theta'_z = C \tan 30°$ where C was given values of 0.85, 0.70, 0.50, 0.35, and 0.20. The intersection heights Y'_z of these rays were then determined. The slopes of the corresponding chief zonal rays in object space were calculated so as to satisfy the tangent condition:

$$\frac{\tan \theta'_z}{\tan \theta_z} = \frac{f}{RF'} \text{ .}$$

These precalculations provided a set of six object space chief ray slopes and a corresponding set of target values for the intersection heights of these rays at the flat image surface. The program was requested to aim the refracted rays through the center of the aperture stop to their respective intersection heights Y'_z. Distortion would be corrected to the degree that the rays in image space were on target.

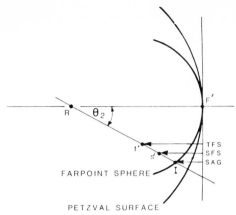

Fig.3. Oblique astigmatic error expressed as distances TFS and SFS. When TFS=SFS=SAG a point image will be formed at 1 and OAE and MOE will be zero

Point I on the far point sphere in Fig. 3 is the ideal image point for the 30° chief edge ray. The distance from the flat image surface to point I measured parallel to the optical axis is the SAG, where:

$$SAG = F'R \left(1 - \sqrt{1-sin^2\theta'}\right) .$$

The program executes a Coddington ray trace to compute the positions on the chief ray of the tangential and sagittal foci but expresses these positions as distances TFS and SFS, also parallel to the optical axis. Astigmatism and power error will be eliminated if TFS = SFS = SAG. Therefore, SAG values were calculated for all of the fractional field angles and the program was requested to set the TFS and SFS values equal to the respective SAG values. Optimization with ACCOS V [8], thus consisted of specifying a rayset containing a chief ray from each of six field angles in object space. Three image errors were defined for each ray, namely, intersection height Y', TFS and SFS. Target values for these 18 image errors were then specified in accordance with the requirements noted above. The optimization subfiles were, thereby, considerably simplified from those used in Reference [1]. The conic constant κ and deformation coefficients d, e, f, and g of both lens surfaces were varied in order to obtain optimization. The equation used to aspherize the surfaces is:

$$z = c\rho^2/ \left[1 + \sqrt{1-(\kappa+1)c^2\rho^2}\right] + d\rho^4+e\rho^6+f\rho^8+g\rho^{10} \qquad where:$$

$$\rho^2 = x^2 + y^2 .$$

The equation represents a surface of revolution with axial curvature c, and its vertex at the origin of the coordinates. When the deformation coefficients are zero, the surface is a conic section of revolution where:

$\kappa < -1$ Hyperboloid

$\kappa = -1$ Paraboloid

$-1 < \kappa < o$ Ellipsoid of revolution about the major axis

$\kappa = o$ Sphere

$\kappa > o$ Ellipsoid of revolution about the minor axis .

Procedures for ray tracing through aspheric surfaces may be found in several references [9-11].

6. Results

The results are summarized in a set of aberration curves for each lens power [12]. The aberration curves were plotted with data calculated at ten fractional field angles in increments of 0.1. The figures also show cross sections of the lenses so that their shapes may be compared.

6.1 -20 Diopter Lenses

Design and aberration data for three -20 diopter best form lenses are indicated by circles in Fig. 2. The aberration curves in Fig. 4 show that the point focal Ostwalt and Wollaston forms have no astigmatic errors (OAE). They also exhibit very small power errors (MOE) since the Petzval surface nearly coincides with the far point sphere of a -20 diopter lens. Distortion at 30° eye rotation, however, is -26.05 and -15.54%, respectively. The minimum distortion form, in Fig. 4c, exhibits -13.89% distortion, an astigmatic error of -4.32 diopters, and a power error of -4.00 diopters. It is clearly not a best form lens.

By aspherizing the surfaces of this form, it was possible to reduce the aberrations to the levels shown in Fig. 4d. Instead of progressively increasing from the center to the edge of the field, the power error peaks zonally at 0.9 full field where it is -0.38 diopters, and distortion reaches 1.41% at 0.8 full field. Astigmatic error is less than 0.36 diopters out to 0.9 full field and then increases rapidly to 1.13 diopters at the edge of the field.

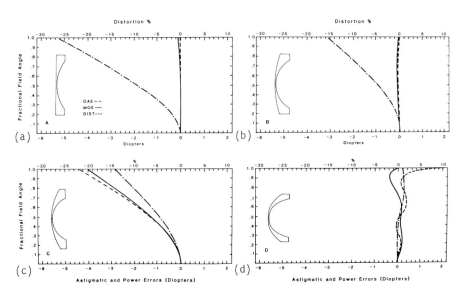

Fig.4. Astigmatism (OAE), power error (MOE), and distortion of -20D lenses: (A) Ostwalt form, (B) Wollaston form, (C) Minimum distortion form, (D) Aspheric form

6.2 -14 Diopter Lenses

Design and aberration data for a similar set of -14 diopter lens forms are
provided in Fig. 5. The point focal forms have distortion of -17.21 and
-8.38%, but are otherwise well corrected. The minimum distortion form exhibits significant amounts of all three aberrations, whereas the optimized
aspheric form of this lens is remarkably well corrected for these aberrations at all zones.

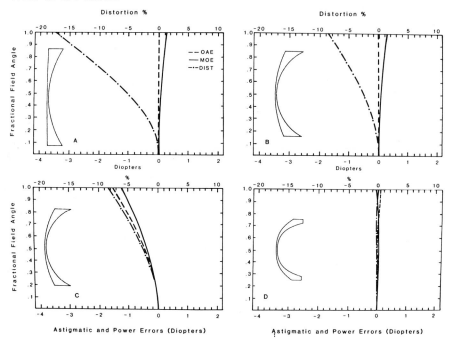

Fig.5. Astigmatism (OAE), power error (MOE) and distortion of -14D lenses:
(A) Ostwalt form, (B) Wollaston form, (C) Minimum distortion form, (D) Aspheric form

6.3 -10 Diopter Lenses

Similar conditions were obtained for the various forms of -10 diopter lenses
(Fig. 6). The aspheric reduces all aberrations to insignificant levels.

6.4 -8 Diopter Lenses

As shown in Fig. 7, the aspheric form provides a significant reduction in
power error and distortion as compared to the spherically surfaced forms.
Astigmatic error, though greater than in the point focal forms, is inappre-
ciable.

6.5 +6 Diopter Lenses

These were the weakest positive lenses studied. The results in Fig. 8, as
expected, show that astigmatic error is corrected in the point focal forms
but power error is somewhat more than -1/3 diopter at 30°. The minimum dis-

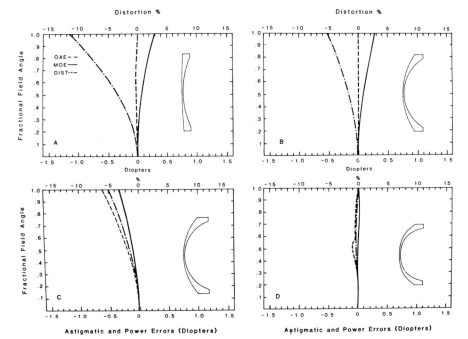

Fig.6. Astigmatism (OAE), power error (MOE) and distortion of -10D lenses: (A) Ostwalt form, (B) Wollaston form, (C) Minimum distortion form, (D) Aspheric form

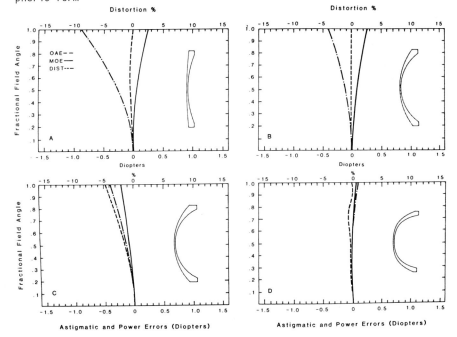

Fig.7. Astigmatism (OAE), power error (MOE), and distortion of -8D lenses: (A) Ostwalt form, (B) Wollaston form, (C) Minimum distortion form, (D) Aspheric form

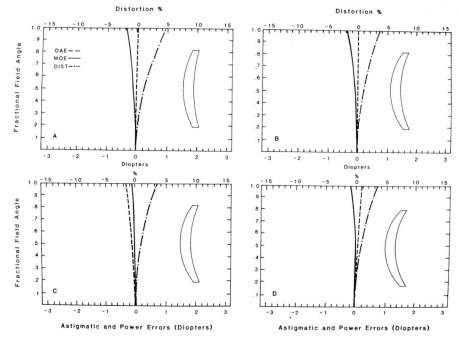

Fig.8. Astigmatism (OAE), power error (MOE), and distortion of +6D lenses:
(A) Ostwalt form, (B) Wollaston form, (C) Minimum distortion form, (D) Aspheric form

tortion form exhibits 0.30 diopters of astigmatism but has very little power
error. Distortion ranges from 4.6 to 3.3% for these forms. The best aspheric form exhibits more astigmatism than the point focal forms but less than
the minimum distortion form, less power error than the point focal forms but
more than the minimum distortion form, and distortion intermediate to the
Ostwalt and Wollaston point focal forms.

6.6 +8 Diopter Lenses

There are no point focal forms for this and the higher power lenses that follow, consequently, the best forms are the minimum astigmatism and minimum
distortion bendings. The data shown in Fig. 9 indicate that the best aspheric form had errors that were intermediate to the other two forms.

6.7 +10 Diopter Lenses

As indicated in Fig. 10, the best aspheric +10 diopter lens form showed reduced astigmatism compared to the other forms, but produced greater power
errors at edge of the field than the minimum astigmatism form. Although
distortion was lowest for the aspheric form, it is greater than desired.

6.8 +14 Diopter Lenses

The results of this, the greatest positive lens studied, are given in Fig.
11. The aspheric form was optimized in accordance with the procedures described in Reference [1], and so differs from all the other optimized forms
described above. The results essentially are well-corrected astigmatic

114

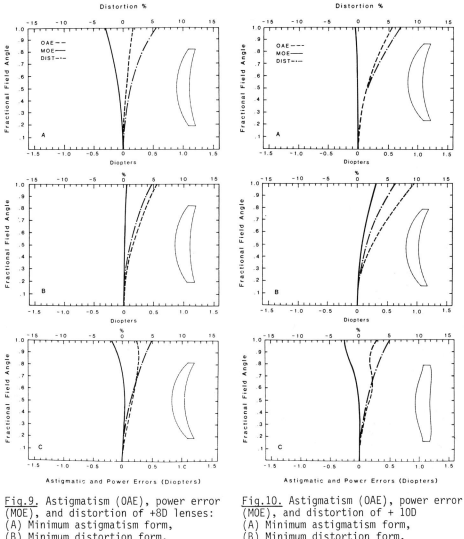

Fig.9. Astigmatism (OAE), power error
(MOE), and distortion of +8D lenses:
(A) Minimum astigmatism form,
(B) Minimum distortion form,
(C) Aspheric form.

Fig.10. Astigmatism (OAE), power error
(MOE), and distortion of + 10D
(A) Minimum astigmatism form,
(B) Minimum distortion form,
(C) Aspheric form

error and distortion but -1.78 D of power error. This lens is further discussed in the next section.

7. Discussion

Successful optimization depends upon properly defining the elements of a
merit function which measures the image defects, applying appropriate weights
to the correction of these defects, and specifying sufficient constructional
parameters of the lens to be modified in order to achieve optimization.

115

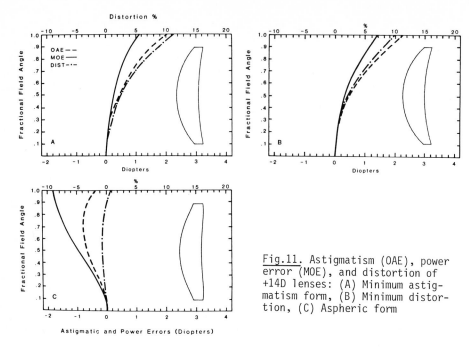

Fig.11. Astigmatism (OAE), power error (MOE), and distortion of +14D lenses: (A) Minimum astigmatism form, (B) Minimum distortion, (C) Aspheric form

Eighteen operands were defined to construct the merit function. As the optimization proceeded, various weights of operands were used. The outcome of optimization (the resultant lens shape), therefore, depended uniquely on these operands and weighting factors.

The maximum number of available constructional parameters, namely, the conic constant and four deformation coefficients of each lens surface, were used to achieve optimization. These proved sufficient for highly correcting the minus lenses, but did not produce equivalent correction in high plus lenses. The resultant deformed conicoids of each lens surface for each lens are:

Vertex Power (D)	cc	Surface Shape
-20	3.532E+00	deformed ellipsoid of revolution about minor axis
	-7.511E-01	deformed ellipsoid of revolution about major axis
-14	2.023E+00	deformed ellipsoid of revolution about minor axis
	2.526E-03	deformed ellipsoid of revolution about minor axis
-10	1.756E+00	deformed ellipsoid of revolution about minor axis
	7.967E-01	deformed ellipsoid of revolution about minor axis
-8	8.336E-01	deformed ellipsoid of revolution about minor axis
	9.357E-o1	deformed ellipsoid of revolution about minor axis
+6	4.572E-03	deformed ellipsoid of revolution about minor axis
	1.158E-03	deformed ellipsoid of revolution about minor axis
+8	-2.400E-03	deformed ellipsoid of revolution about major axis
	-9.205E-02	deformed ellipsoid of revolution about major axis
+10	-3.270E-01	deformed ellipsoid of revolution about major axis
	-1.621E+01	deformed hyperboloid
+14	-3.665E-01	deformed ellipsoid of revolution about major axis
	-1.568E+01	deformed hyperboloid

116

Except for the second surface of the -20 diopter lens, both surfaces of all negative lenses became deformed ellipsoids of revolution about the minor axis. The +6 and +8 diopter lenses have conic constants that are very nearly zero and might be considered to be deformed spheres. This is true of some surfaces of the negative lenses. The high-powered positive lenses tended to assume front surfaces that are deformed ellipsoids about the major axis and deformed hyperboloidal second surfaces.

Although the +14 diopter lens reported here was optimized according to Reference [1] methods, it may be useful to compare its performance at 30° with the +14 diopter aspheric lenses described by FOWLER [13], and a lenticular cataract lens by DAVID and FERNALD [14].

Design	OAE (D)	MOE (D)	Dist. (%)
Katz	-0.30	-1.78	0.91
Fowler ellipsoid	-0.31	-1.07	6.23
Fowle paraboloid	-2.64	-3.54	0.44
David and Fernald	+0.71	-0.13	8.45

In none of these designs are all three aberrations corrected.

Conclusion

It is possible to correct fully high minus lenses by using deformed ellipsoids of revolution on both lens surfaces. The resultant shapes have rather large vaults which may present cosmetic and mechanical impediments to their use. It does not appear feasible to correct all aberrations of strong positive lenses despite the use of deformed conicoids on both surfaces. However, further work with moderate positive lenses is indicated.

References

1   M. Katz: Applied Optics 21, 2982 (1982)
2   M. von Rohr and H. Boegehold: *Das Brillenglass als Optisches instrument* (Springer, Berlin, 1934), pp. 110-140
3   O. Henker: *Introduction to the Theory of Spectacles*, R. Kanthack, translator (Jena School of Optics, 1924)
4   M. Jalie: *The Principles of Ophthalmic Lenses* (The Association of Dispensing Opticians, London, 1977)
5   A.G. Bennett and D.F. Edgar,"Spectacle Lens Design and Performance," a 12-part series of articles in The Optician (22 July 1979 to 10 Oct 1980)
6   E.W. Bechtold: Am. J. Optom. Arch. Am. Acad. Optom. 35 (1958)
7   G.A.Fry: Am. J. Optom. Arch. Am. Acad. Optom. 55, 238 (1978)
8   ACCOS V Manual, Scientific Calculations, Inc. (1976)
9   W.J. Smith: *Modern Optical Engineering* (McGraw-Hill, New York, 1966), pp. 247-262
10 R. Kingslake: *Lens Design and Fundamentals* (Academic, New York, 1978), pp. 19-36
11 G. Smith and L. Bailey, "Aspheric Spectacle Lenses--Design and Performance," The Optician Part 1 (10 Apr 1981); Part 1 (8 May 1981); Part 3 (19 June 1981)
12 Tables of data from which the various figures were constructed are available from the author.
13 C. Fowler, Aspheric Lenses, The Optician (13 August 1982)
14 J.K. David and H.G. Fernold, Ophthalmic aspheric lens series, US Patent No. 3,169,247 (1965)

A New Sphero-Cylinder Trial Lens for Aphakic Correction and Its Clinical Application

Keiichiro Kato, Masanori Sakai, Teruo Komine

Department of Ophthalmology, Fukushima Medical College, Fukushima-ken, Japan

Haruo Oda

Nippon Kogaku K.K., Tokyo, Japan

Introduction

Recently, a great change has occurred in the methodology of aphakic correction with the use of the intraocular lens or soft contact lens for extended wear. Therefore, we could not find many recent reports concerning spectacle-correction after cataract extraction. But at least in Japan, we are more likely to use spectacles for aphakic correction. Aphakic correction with spectacles is still an easier and safer method in spite of several optical disadvantages.

From a clinical viewpoint, one of the most important problems in aphakic correction seems to be the dioptric error between refracted and finished lenses, especially in sphero-cylinder combination.

Consequently, we often found patients whose visual acuity through the finished spectacle lenses was not the same compared with that of trial lenses composed of individual spheres and cylinders. For aphakic correction, we are accustomed to use the trial lenses designed with spheric curves such as biconvex or meniscus and then actually to prescribe eye glasses designed with aspheric curve. Serious errors of power between two conditions can easily result.

Trial Lens Design and Construction

For the purpose of making spectacle prescriptions more precise, we prepared new trial lenses made of the same material and design as aspherical sphere and cylinder combinations. These trial lenses are based on the design of Nikon aspherical lenticular lenses for aphakic spectacles made of the plastic CR 39. We cut the lens 40 mm in diameter to keep the effective area as large as possible and set them in proper frames (Fig. 1a). A special trial frame had to be prepared for adjusting to these lenses.

Astigmatism of aspherical trial lenses is very small. Figure 1b reveals the astigmatism of +12 aspherical trial lens compared with a corresponding meniscus lens.

We called this set of new trial lenses, aspherical sphero-cylinder trial lenses or combination sphere and cylinder aspherical trial lenses, aspherical CSC trial lens for short.

Table 1 shows the standard composition and central thickness of each lens, ranging in powers from +8 to +16 D for spheres and to +2 for cylinders, in 0.5 D steps.

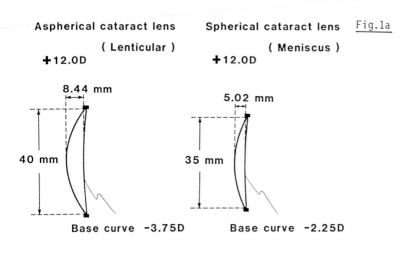

Aspherical cataract lens   Spherical cataract lens   Fig.1a

(Lenticular)   (Meniscus)

+12.0D   +12.0D

8.44 mm   5.02 mm

40 mm   35 mm

Base curve -3.75D   Base curve -2.25D

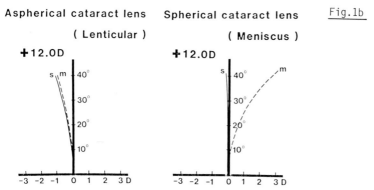

Aspherical cataract lens   Spherical cataract lens   Fig.1b

(Lenticular)   (Meniscus)

+12.0D   +12.0D

Fig. 1 a and b Comparison of trial lenses (design and their astigmatism)

As these lenses were produced by precise mechanical means, the errors are
minimal for each of the following conditions: within 0.1 D for the power,
0.5 mm for geometric center, and 1 deg for the cylindrical axis.

Sphere and cylinder ranging to +3 D, frosted or Chavasse lens for the occluder, and also a colored lens for absorbing excess light were available
as additional lenses. Color lenses were dyed and numbered according to their
density.

Clinical Results and Comment

The following procedure could be recommended as one of the practical uses
for this trial set.

1. Adjust the vertex of sphere and interpupillary distance.
2. Select the suitable power for sphere and additional cylinder.
3. Confirm the visual acuity and recorrect with aspherical CSC trial lens.
4. Readjust the interpupillary distance in binocular correction by prism
   neutralization or with maddox rod.
5. Evaluate the new visual condition with a pair of the selected lenses.

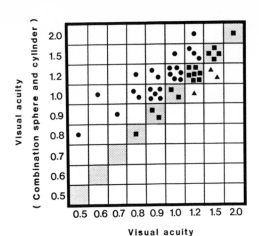

Fig. 2 Visual acuity corrected by aspherical CSC compared with pair of meniscus trial lenses

If these several procedures are satisfactory, a final prescription can be done.

For evaluating the difference of visual acuity between two trial procedures, 44 uneventful aphakic eyes were studied. Figure 2 shows the visual acuity corrected by aspherical CSC compared with duplicated meniscus trial lenses on these cases. Details of the results are as follows.

The vision was improved in 24 (55%) eyes, for 10 eyes (23%) over two·steps on the Landolt chart; it remained unchanged in 17 eyes (39%) and worse in 3 (7%) of the eyes corrected by aspherical CSC trial lenses.

As we could not find a definite rule in the errors of power between each trial, several examples in practice are compared with the calculated results (Fig. 3, a-c). In these trials, we had to keep the vertex distance to 10 mm because of the thickness of aspherical CSC trial lenses.

Each case in Fig. 3 is tested for the best vision, the power of the trial lenses which were used for the final corrections and also the estimated power from the results of duplicated meniscus correction is listed.

From these results, we can easily recognize that there may be unexpected errors between the power of duplicated trial lenses and that of finished spectacles, especially in power of sphere. Several factors are involved in this problem.
1. Difference of lens design (material, thickness, base-curve, asphericity)
2. Position of additional cylinder
3. Decentration effect
4. Tilting effect ·

Decentration and tilting of lens seems to exert a great influence on the errors of power; however, it is impossible to estimate precisely these factors, especially in duplicated trial lenses.

One of the reasons that we made the single trial lens was to minimize errors of power as much as possible.

EYE 1   S.T.   RIGHT EYE

A. Spherical cataract lens

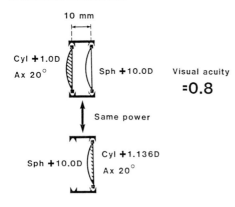

Cyl +1.0D
Ax 20°   Sph +10.0D   Visual acuity
                               =0.8

Same power

Sph +10.0D   Cyl +1.136D
               Ax 20°

B. Aspherical cataract lens

Sph +11.0D   Cyl +1.5D   Visual acuity
               Ax 20°   =1.2

EYE 2   S.T.   LEFT EYE

A. Spherical cataract lens

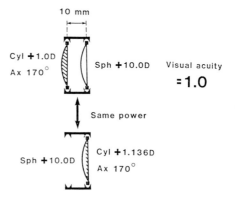

Cyl +1.0D
Ax 170°   Sph +10.0D   Visual acuity
                              =1.0

Same power

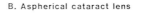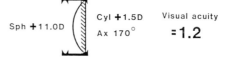

Sph +10.0D   Cyl +1.136D
               Ax 170°

B. Aspherical cataract lens

Sph +11.0D   Cyl +1.5D   Visual acuity
               Ax 170°   =1.2

EYE 4   S.A.   RIGHT EYE

A. Spherical cataract lens

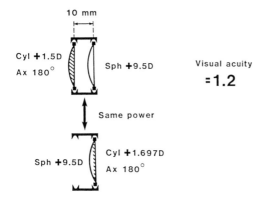

Cyl +1.5D
Ax 180°   Sph +9.5D   Visual acuity
                               =1.2

Same power

Sph +9.5D   Cyl +1.697D
               Ax 180°

B. Aspherical cataract lens

Sph +10.5D   Cyl +2.0D   Visual acuity
               Ax 180°   =1.2

Fig. 3a-c Examples in practice
of (A) the power of a pair of
meniscus lenses to obtain best
vision and estimated power when
replaced with single lens, compared with (B) power of aspherical CSC trial lens for best
visual acuity

Table 1 Makeup of aspherical CSC trial lenses and central thickness of each lens.   (mm)

	Sph	Cyl+1.0D	Cyl+1.5D	Cyl+2.0D
+ 8.0D	6.3	6.8	6.9	7.3
+ 8.5D	6.8	6.9	7.3	7.4
+ 9.0D	6.8	7.3	7.4	7.8
+ 9.5D	7.3	7.4	7.8	7.9
+10.0D	7.3	7.8	7.9	8.3
+10.5D	7.8	7.9	8.3	8.4
+11.0D	7.8	8.3	8.4	8.8
+11.5D	8.3	8.4	8.8	8.9
+12.0D	8.3	8.8	8.9	9.2
+12.5D	8.8	8.9	9.2	9.3
+13.0D	8.8	9.2	9.3	9.7
+13.5D	9.2	9.3	9.7	9.8
+14.0D	9.2	9.7	9.8	10.2
+14.5D	9.7	9.8	10.2	8.8
+15.0D	9.7	10.2	8.8	9.2
+15.5D	10.2	8.8	9.2	9.3
+16.0D	10.2	9.2	9.3	9.6

CSC: Combination Sphere and Cylinder

Outer diameter: 40.5 mm

Effective diameter: 39.3 mm

Table 2 Difference of brightness between both eyes in monocular aphakia

	Visual acuity	Colour−density required for same brightness
I.W.	R) 1.2X+8.0D L) 1.2X−1.5D	20 %
T.M.	R) 1.0X+9.0D L) 1.0 (n.c.)	15 %
E.E.	R) 0.7X+6.5D≃cyl+4.5DA20° L) 0.3X−5.5D (Cat. capsularis post.)	50 %
H.T.	R) 0.9X+12.0D≃cyl+1.25DA165° L) 0.6X+0.25D≃cyl−0.75DA135° (Cat. capsularis post.)	75 %
K.K.	R) 1.2X+6.0D≃cyl+3.0DA165° L) 0.6X−5.5D (Cat. nuclearis)	40 %

Most aphakic patients complain about glare or brightness due to trans-mitted light. Dyeing the surface of plastic lenses is the best method to eliminate this glare. For choosing the suitable density of dyeing, clini-cal experiment was made on dyes with various densities of NIKON Brown. Table 2 shows the results of the experiment.

In the first and second cases whose fellow eyes are almost normal, we could equate for brightness after setting 15% or 20% NIKON Brown lenses in front of the aphakic eye. On the other hand, in the third and fourth patients with monocular posterior capsular cataract, there seemed to be great differences of brightness, such as 50% or 75%, between both eyes. That is to say, most light may be absorbed by the cataractous lens in these eyes. Additional color lenses that are available should be used to decide the tint of the finished spectacle lens prescription.

Corneal Specular Microscopy Using a Contact Lens

Yoshinori Ohno, Noboru Kaneko, Tatsuya Deguchi, and Kimiharu S. Noyori

Saitama Medical School, Department of Ophthalmology, Saitama-ken, Japan

Introduction

Corneal specular microscopy has been progressing rapidly [1]. Most of the newer techniques utilize a water immersion microscope objective with direct contact with the cornea [2]. But the cornea is easily scratched, and the technique is rather difficult to use. We have reported non-contact specular microscopy using a high resolution objective elsewhere [3]. The latter technique has some merits; e.g. it diminishes light reflections from the corneal surface, and produces a sharper image than can be obtained compared to the non-contact type. However, in spite of these advantages it must be emphasized that non-contact viewing is technically simpler from the examiner's point of view.

For these reasons, a combination of both techniques would be superior in obtaining sharp pictures with ease of operation for photographing corneal endothelium. This report concerns a specular microscopic technique utilizing a specially designed contact lens in combination with a newly designed high resolution objective.

Construction of the System

The system is composed of a slit lamp camera with high resolution objective (10 X) and a single element contact lens as shown in Fig. 1.

The objective has a focal length of 28 mm, f: 1.4, and is made from a 12-element glass lens. It has a very high resolution power compared to other objectives as shown in Table 1.

The contact lens is a single-element glass lens which has an aspherical surface on the front. This lens is similar in appearance to those lenses used for fundus photography [4] and is shown in Fig. 2A and B. Previously, we experimented with various types of contact lenses (Fig. 3A). Each lens is supported by an applanation tonometer stand [5] and it is touched lightly to the anesthetized cornea. However, none of the lenses gave a satisfactory result mainly due to the difficulty of placing the lens in the position representing the conjugate point of the axes of both the photographic and illuminating systems.

The contact lens is held by hand in the same manner as for observing the fundus, and a methylcellulose solution is used prior to placing the lens on the cornea. The angle between the optical axes of the photographic and illuminating system is set at approximately 60°, which gave us maximum brightness by the mirror effect of corneal endothelium.

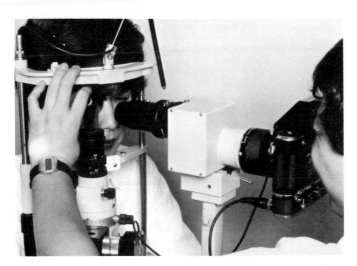

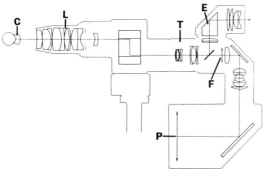

Fig. 1A The slit lamp camera
with the contact lens in use

Fig. 1B Schema of the system.
C, contact lens; L, 10X objective lens; T, 2X image converter lens; E, view finder;
F, 35 mm film plane; P, po-
laroid film plane

Table 1 Comparison of resolving power of various objective lenses. Under-
lined resolution value is the objective used in the present study.

	magnification (focal length)	F. stop	N.A.	resolution
New lenses	5X (50 mm)	F:2.6	0.16	550 lines/mm
	10X (28 mm)	F:1.4	0.32	1,150 lines/mm
Slit-lamp object.	3.5X	F:5.6	0.07	250 lines/mm
Photographic lens	10X (35 mm)	F:4.5	0.1	360 lines/mm

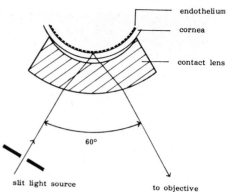

Fig. 2A General appearance of the contact lens

slit light source

to objective

Fig. 2B Schema of slit beam reflected back from the endothelium through the contact lens

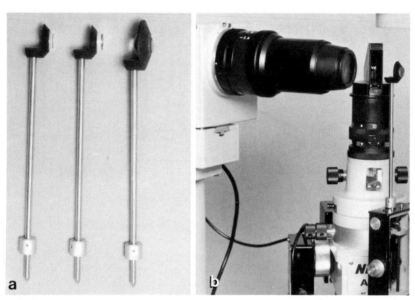

Fig. 3A Trial models of contact lens

Fig. 3B The contact lens supported by an applanation tonometer stand

Results

Figure 4 shows several photographs taken by this system compared with photographs taken without a contact lens (non-contact specular microscopy). It is obvious that photographs taken with the lens are superior.

Image magnification on the film is 20X with combined use of a 2X converter lens placed in the photo-taking system. The final image magnification on the print is about 400X, by which time the size of the endothelial cells can be easily measured.

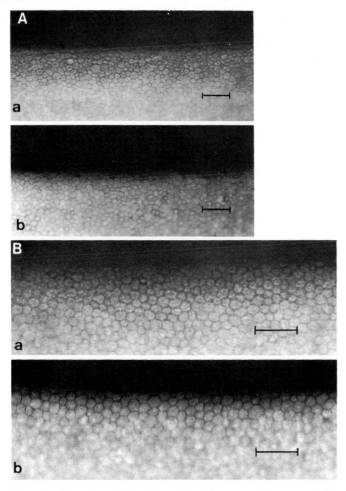

Fig. 4A Normal endothelium of a 19-year-old woman. Cell density: 2600 cells per mm^2; cell size: 316.6 ± 109.3 µm^2; bar gauge: 100 µm (x 100). a) photographed without contact lens; b) photographed with contact lens.

Fig. 4B Corneal endothelium of a 29-year-old man, with an old perforative injury. Cell density: 1933 cells/mm^2; cell size: 554.2 ± 102.1 µm^2; bar gauge: 100 µm (x 150). a) photographed without contact lens; b) photographed with contact lens.

One application of this system is the exact measurement of the corneal endothelium by using a digitizer [6], which can sum up the total area of a particular section (Fig. 5). This method can also be applied for comparison of the size differences of the endothelium by age or after surgery. Figure 6 shows that the size of the corneal endothelium cells becomes proportionately wider as a function of age. It is now recognized that the size of the endothelium cells becomes larger after corneal injury or after corneal incisions are surgically made [7].

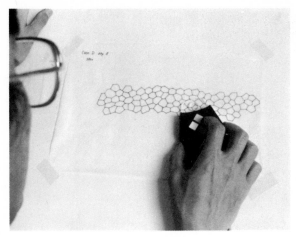

Fig. 5 Drawing the cell outline with a cathode on the table of the digitizer.

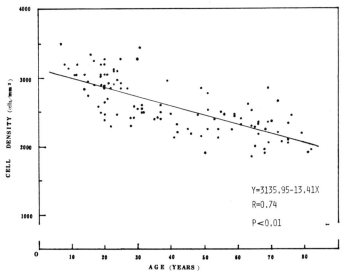

Y=3135.95-13.41X

R=0.74

P<0.01

Fig. 6 Cell density of normal endothelium vs. age, N = 106 corneas.

Discussion

Specular microscopy is based on the principle that corneal endothelium can be photographed when a slit beam is projected to the endothelium and reflected back like a mirrored surface. Therefore, in order to obtain a high quality image, the axes of the illuminating and photo-taking system must be arranged to get maximum brightness from the endothelium, but at the same time avoiding the light reflection from the corneal surface.

In contact-type specular microscopy, surface reflection can be diminished by separating the paths of the illuminating and photographic otpics with a

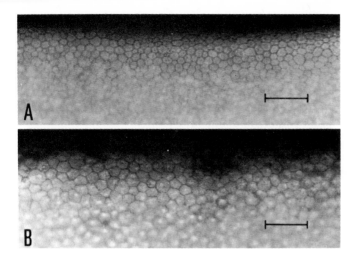

Fig. 7A Normal endothelium of a 27-year-old woman. Cell density: 2844 cells per mm^2; cell size: 310.6 ± 60.8 μm^2; bar gauge: 100 μm (x 150).

Fig. 7B Corneal endothelium of a 23-year-old man. He suffered a performation on his cornea 3 years ago; cataract operation was performed 7 months ago. Cell density: 1290 cell/mm^2; cell size: 612.9 ± 154.2 μm^2; bar gauge: 100 μm (x 150).

microscope objective, which is placed on the corneal surface. However, it is rather difficult to avoid the corneal reflection in the case of non-contact microscopy, due to direct reflection from the corneal surface. By incorporating a contact lens, the light reflection can be diminished similar to contact-type microscopy where a lens is on the cornea. Furthermore, manipulation of the device becomes easier since a photographic unit is independent of the contact lens.

Slit illumination is important in obtaining higher contrast of the object, and it also depends primarily on the width of the slit beam, which in turn limits the brightness of the image. In our experiment, the optimum width of a slit beam is 0.5 mm for a reasonable compromise between image contrast and brightness.

Various applications of corneal specular microscopy have been reported, including photography of corneal stroma and epithelium [8]. However, in our experiment, photography of the latter layers of the cornea has not yet been accomplished. Improvement of the contact lens and the microscope unit would first be necessary for these types of observations.

Summary

The object of this method was to obtain larger and sharper photographs of the corneal endothelium using a specially designed contact lens combined with a high resolution objective. The image magnification on the film is 20X and can be enlarged up to 400X in print, by which means exact measurement of the corneal endothelial cells can be made.

References

1 L.E. Lohman, G.N. Rao, and J.A. Aquavella: Am. J. Ophthalmol. 92, 43 (1981)
2 W.M. Bourne and H.E. Kaufman: Am. J. Ophthalmol. 81, 319 (1976)
3 K. Yabe, Y. Ohno, and K. Chiyoda: Acta. Soc. Ophthalmol. Jpn. 84, 1239 (1980)
4 K.S. Noyori, K. Chino, T. Deguchi, and K. Amano: Ophthalmology Jpn. 24, 481 (1982)
5 R.A. Moses: *Adler's Physiology of the Eye*, 7th ed., pp. 242-247 (C.V. Mosby, St. Louis, 1981)
6 G.O. Waring III, M.A. Krohn, G.E. Ford, R.R. Harris, and L.S. Rosenblatt: Arch. Ophthalmol. 98, 848 (1980)
7 G.N. Rao, E.D. Shaw, E. Arthur, and J.V. Aquavella: Arch. Ophthalmol. 96, 2027 (1978)
8 C.J. Koester, C.W. Roberts, A. Donn, and F.B. Hoefle: Ophthalmology 87, 849 (1980)

Calculatinged Tear Volumes Under Thin Hydrogel Contact Lenses

David D. Michaels and Barry A. Weissman

Department of Ophthalmology and Jules Stein Eye Institute, UCLA School of
Medicine, Los Angeles, CA, USA

Introduction

The effective power of a hard contact lens on the eye is generally considered
to be the sum of the powers of the plastic lens, and the tear fluid "lens"
separating it from the cornea [1]. Hydrogel or soft lenses on the other hand
presumably conform to the anterior corneal surface. Nevertheless it is known
that hydrogel lenses ordered on the basis of keratometry readings (instead of
over-refracting) may not adequately correct the eye. For example they tend
to undercorrect aphakia but not myopia. It has therefore been proposed that
because soft lenses are pliable, they bend or flex. Such flexure can not
only alter effective power but create a tear lens as well.

Several theories have been proposed to describe the optical and mechanical
effects of hydrogel lens flexure in situ. WEISSMAN and ZISMAN [2,3] mathematically described seven models available in 1981 and developed a computer
program to predict the consequences of each. They tested a series of standard central thickness (about 0.15 mm) hydrogel lenses and concluded that an
optically significant tear lens exists of about 10 μl. CHASTON and FATT [4]
used Weissman and Zisman's analysis and developed an independent hypothesis
incorporating their own clinical data for plus lenses. More recently, HOLDEN
and ZANTOS [5] suggested that the flexed back surface of a soft lens comes
into complete alignment with the anterior corneal surface rather than allowing a "tear lens" to form.

We herein add these two additional models to Weissman and Zisman's original program and also evaluate their effect with more modern thin soft contact lenses.

Materials and Methods

Table 1 shows mathematical statements which predict the resultant flexed
back curvature (r_2') of a lens for each previously described hypothesis.
Table 2 shows the two additional models, rewritten mathematically, to predict r_2'.

Fifteen lens/eye systems were studied. Modern thin (all about .06 mm in
central thickness) soft contact lenses were used for this study. Table 3
shows manufacturers' parameters, and each input measurement described below.
These lens systems were analyzed by a computer program (available from the
author) run on a VAX-11/780 main frame computer; fourteen lens systems generated usable data.

Initial keratometry for each eye studied was obtained with a Bausch & Lomb
keratometer calibrated with steel spheres prior to use. Lenses were inserted

TABLE 1. Mathematical statements of flexure hypotheses.

Hypothesis	Author (Date)	Statement
1. Equal change	Kaplan (1966) Sarver (1976)	$r_2' = r_2 - r_1 + r_1'$
2. Percentage change	Strachan (1973)	$r_2' = (r_2 \cdot r_1')/r_1$
3. Constant volume	Bennett (1976)	$r_2' = \dfrac{\left(3\left(\dfrac{h_1'^2}{3}(3r_1' - h_1') + y^2 k - \dfrac{V}{\pi}\right)\Big/ h_2'^2\right) + h_2'}{3}$
4. Constant arc length	Wallace-Williams & Magabilen (1972); Holden et al. (1976)	Simplifies to the percentage change hypothesis exactly in the case of a plus lens and closely in the case of a minus lens
5. Invariant normals	Wallace-Williams & Magabilen (1972); Holden et al. (1976)	$r_2' = \dfrac{y'^2 + [(r_1'^2 - y'^2)^{\frac{1}{2}} - (r_1' - t_c)]^2 - t_y^2}{2t_y - 2[(r_1'^2 - y'^2)^{\frac{1}{2}} - (r_1' - t_c)]}$ where $y' \cong \left(\dfrac{r_1'}{r_1}\right)^2 \cdot y$
6. Constant sag change	Smith (1978)	$r_2' = \dfrac{(r_1' - r_1)}{(r_1^2/r_2^2)} + r_2$
7. Power	Weissman & Zisman (1979)	$r_2' = \dfrac{(1 - n) + 0.3375}{P_e - \dfrac{(n - 1)/r_1'}{1 - \left\{\dfrac{t_c(n - 1)}{nr_1'}\right\}} - \dfrac{0.3375}{r_c}}$

Table 2 Two additional statements of recent innovation

| 8. Chaston & Fatt | Chaston & Fatt (1980) | $r_2' = r_2 \ (r_1'/r_1)^2$ |
| 9. Alignment | Holden & Zantos (1981) | $r_2' =$ anterior corneal curvature |

and allowed to settle for at least 15 minutes. Overkeratometry was performed with the same instrument by the same observer in each case. The effective power (P_e) of each lens on the eye was found by subjective over-refraction added to refractive error. The lens was then removed, rehydrated for at least 15 minutes, and its unflexed base curve (r_2) measured by the radiuscopic method of PARELHOFF and WEISSMAN [6] with a standard American Optical radiuscope. Lens central thickness (t_c) was measured on the same instrument by the direct method also described by Parelhoff and Weissman among others. Unflexed lens back vertex power was measured by a standard lensometer according to the technique of SARVER et al. [7].

Results

No single model appears to account fully for the effective powers of flexible soft contact lenses on the eye. Table 4 shows the various hypotheses which in each lens system example best predicted lens power behavior (probably the best way to select a realistic hypothesis). Also shown are the difference between predicted power under that hypothesis and the measured effective power, and the predicted tear volume in each case. The "Percent Change" hypothesis and the "Alignment" model each were best in three of fourteen cases. The "Invariant Normals" hypothesis gave poorest results (in agreement with earlier work of Weissman and Zisman), followed closely by that of CHASTON and FATT [4]. Surprisingly, the "Equal Change" hypothesis also gave poor results in three lens systems. Mean error in power prediction, however, for the best model in each case, was about 1/8 of a diopter. The predicted tear volume average was about 4 µl.

The "Power" hypothesis offers good comparison as it is based on correct effective power. This model predicts an average tear layer of 9 µl, whereas the "Alignment" hypothesis suggests a trapped tear volume less than 0.5 µl in all cases (Table 5).

Conclusions

Computer models are limited by the assumption of spherical surfaces for all calculations [2,3]. This assumption is reasonable for unflexed parameters, but perhaps wanting for in situ flexed measurements. Nonetheless, it is the best available for those central areas of the lens and cornea which participate in optical refraction. Accepting the assumption that these systems form spherical surfaces with flexure, our analysis suggests that some but not all modern thin soft contact lenses may closely align with the anterior corneal surface producing a tear layer of negligible power and minimum volume. Many lens systems, however, do have some tear lens power and volume (about .15 diopters and 5 to 10 µl respectively). In no case does the formed tear volume contribute significantly to the oxygenation of the cornea. The refractive change may, in some cases, alter the final acuity. Clinicians may find this concept useful in fitting and in evaluating the optical response of patients to hydrogel contact lenses.

Table 3 Thin lens flexure input data

lens #	labled parameters	back vertex power (D)	base curve r₂ (mm)	thickness t_c (mm)	index	effective power (D)	K (mm)	OverK (mm)
1	8.4/-2.00/14	-2.00	8.0	.07	1.43	-1.75	7.80/7.63	8.08/8.04
2	8.4/-1.75/14	-1.75	8.3	.09	1.43	-2.00	7.78/7.63	8.08/7.87
3	8.6/-3.00/13.8	-4.00	8.4	.08	1.43	-4.00	7.63/7.67	8.18/8.13
4	8.9/-2.00/14.5	-2.75	12.55	.06	1.43	-2.00	7.54/7.58	8.04/7.90
5	8.9/-2.00/14.5	-2.25	8.99	.05	1.43	-2.00	7.58/7.50	8.04/7.90
6	8.3/-4.75/13.5	-4.75	9.85	.05	1.43	-5.00	7.26/7.07	7.94/7.71
7	8.6/-5.25/13.5	-5.25	12.55	.04	1.43	-5.25	7.63/7.50	8.44/8.28
8	8.3/-4.75/13.5	-5.00	9.85	.04	1.43	-5.25	7.34/7.18	8.04/7.85
9	8.6/-5.00/13.5	-5.00	12.55	.045	1.43	-5.00	7.63/7.50	8.28/8.28
10	8.6/-1.00/13.5	-1.00	10.3	.04	1.43	-1.00	7.71/7.63	7.85/7.76
11	8.5/-3.25/14.0	-3.75	11.7	.04	1.43	-3.25	7.58/7.38	8.33/7.94
12	8.6/-1.00/13.5	-1.00	10.3	.036	1.43	-1.25	7.71/7.67	7.85/7.76
13	8.3/-1.50/13.5	-1.75	9.85	.07	1.43	-2.00	7.22/7.18	7.46/7.46
14	8.5/-4.75/14	-4.50	11.7	.04	1.43	-2.75	7.58/7.42	8.08/7.90
15	8.3/-1.50/13.5	-1.75	9.85	.06	1.43	-2.25	7.22/7.14	7.46/7.46

Table 4

lens #	hypothesis	power difference (diopters)	predicted tear volume (ul)
1	Const. vol.	.174	1.51
2	Eq. change	.223	2.00
3	Eq. change	.299	10.66
4	Const. vol.	.012	3.28
5	Const. vol.	.078	4.15
6	Alignment	.045	0.33
7	Alignment	.081	0.34
8	Perct. change	.196	14.43
9	Alignment	.172	0.34
10	Perct. change	.019	2.61
11	Const. vol.	.386	6.41
12	Eq. change	.267	6.50
13	Perct. change	.015	3.23
14	no usable data generated		
15	Perct. change	.083	1.91
mean values		.15	4.12

Table 5 Tear volumes predicted by power and alignment models

lens #	POWER MODEL (ul)	ALIGNMENT MODEL (ul)
1	8.48	0.34
2	8.19	0.34
3	19.89	0.34
4	13.43	0.34
5	16.19	0.34
6	1.54	0.34
7	1.71	0.34
8	7.49	0.34
9	4.92	0.34
10	3.15	0.34
11	21.11	0.34
12	14.28	0.34
13	2.71	0.34
14	no usable data generated	
15	4.78	0.34
mean values	9.13	0.34

References

1   D.D. Michaels: *Visual Optics and Refraction*, 2nd ed. (C.V. Mosby, St. Louis 1980), p. 641
2   B.A. Weissman and F. Zisman, Am. J. Optom. and Physiol. Optics 56, 727 (1979)
3   B.A. Weissman and F. Zisman: Am. J. Optom. and Physiol. Optics 58, 2 (1981)
4   J. Chaston and I. Fatt: The Optician, p. 12 (Nov. 7, 1980)
5   B.A. Holden and S. Zantos: Am. J. Optom. and Physiol. Optics 58, 139 (1981)
6   E.S. Parelhoff and B.A. Weissman: Using a radiuscope to measure base curves of soft lenses, in preparation.
7   M.D. Sarver et al.: Am. J. Optom. and Arch. Am. Acad. Optom. 50, 195 (1972)
8   M.M. Kaplan: Optom. Weekly 57, 22 (1966)
9   M.D. Sarver: Optician 171m 26 (1976)
10 J.P.F. Strachan: Austr. J. Optom. 56, 25 (1973)
11 A.G. Bennett: Ophthalmic Optician 16, 939 (1976)
12 Wallace-Williams and Magabilen: quoted by Weissman and Zisman (1979)
13 B.A. Holden et al.: Austr. J. Optom. 59, 117 (1976)
14 F.D. Smith: (personal communication)

Part 4

Electronic Visualization
of the Fundus

An Overview of the Scanning Laser Ophthalmoscope

Robert H. Webb

Eye Research Institute of Retina Foundation, Boston, MA 02114, USA

The Scanning Laser Ophthalmoscope (SLO) was conceived originally as a recording ophthalmoscope requiring substantially less light than existing fundus cameras. Since then it has demonstrated a wide range of applicability as a diagnostic tool as well as its original ophthalmoscopic function. This paper describes some of the areas in which the SLO is currently being used.

Modern indirect ophthalmoscopes subject patients to retinal irradiances of as much as 200 mW/cm^2, and examinations may last for tens of minutes [1]. Fundus cameras are somewhat less distressing, using irradiances in the 10 mW/cm^2 range, for shorter times. The Scanning Laser Ophthalmoscope [2] requires only 100 uW/cm^2 (0.1 mW/cm^2) to achieve color imaging of a quality close to that of 35 mm cameras (Fig. 1). The increased efficiency of the SLO's "inverted" optical system [3] thus makes possible both ophthalmoscopy and fundus photography with 1% or less of the irradiance in current use. Besides increased patient comfort and safety, this provides the physician with an opportunity to observe a retina for an extended time and to consult with others about an image easily visible to more than one observer. Since the image appears on a TV monitor, even a room full of students can simultaneously look at the patient's eye. Video tapes record this view 60 times per sec. The original search for a low-light level recording ophthalmoscope has been successful.

Another consequence of the efficient optics of the SLO is that fluorescein angiography may be performed with the same retinal irradiance as used for ophthalmoscopy (100 uW/cm^2), and with 10% or less of the normal amount of injected fluorescein (1 cc Na-Fluorescein) (Fig. 2). This contrasts with the 10 w/cm^2 x 0.1 sec and 10 cc Na-Fluorescein of current practice. Because the excitation light is blue, this irradiance level is particularly important. Because some 8% of patients react adversely to fluorescein injection, the dosage is of concern.

The use of laser illumination calls attention to the appearance of the retina in strictly monochromatic light. Currently we use 633 nm (red) and 568 nm (yellow) for color imaging (the two lasers are used simultaneously), and 502 nm to excite the fluorescein in angiography. The coherence effects of laser light which result in speckle and other interference effects do not appear in the SLO, since the retina is a focal surface for the laser beam.

The SLO derives its image from the laser raster (the pattern of the swept beam) which falls on the patient's retina. If we turn off the laser beam briefly, a dark spot appears in the raster. The patient perceives this dark spot if it falls on functional retina, and the observer perceives it no matter where it falls, since no light reaches the detector when the beam is off. The fixation cross (in Fig. 1) is generated by such modulation of the laser.

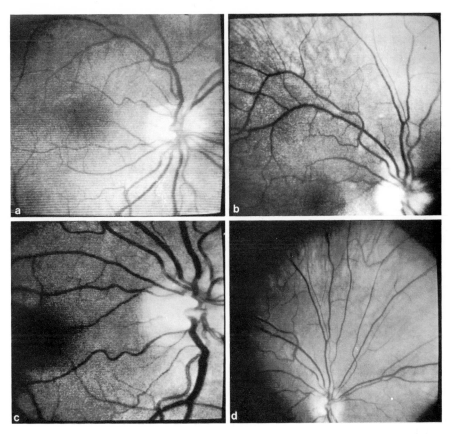

Fig. 1 Photographs of TV monitor showing fundus views of a) disc and macula; b) supero-temporal to disc; c) magnified view of disc and macula, and d) supero-nasal to disc.

The patient sees the cross, and the image shows that it is, in fact, in the macula. Figure 3 shows a more elaborate modulation, in which text appears in the raster. The patient is reading this, and the part of the retina used is apparent on the TV monitor [4].

If we position a dark area in a scotoma (or, in Fig. 4, in the disc), the patient does not perceive it. As this stimulus is moved into the seeing retina, the patient indicates the instant of first seeing it, and its coordinates there are recorded. From this is generated a map of the scotoma--corrected for fixation errors and relateable to specific retinal features. This high—precision projection perimetry, which we designate "scotometry," takes but a few minutes per isopter [5].

The Scanning Laser Ophthalmoscope, in its simplest form, provides fundus photography and projection perimetry with low light levels, low fluorescein dosage and direct visualization of retinal stimuli. As this instrument moves on to more advanced forms it will realize the promise of image processing, interactive beam control and more sophisticated psychophysics.

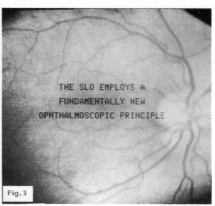

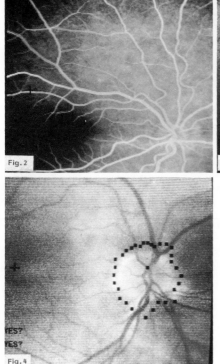

Fig.2 Fluorescein angiogram

Fig.3 Text projected on the retina,
being read by subject

Fig.4 Scotometry map of the optic
nerve head

References

1   F.C. Delori, M.A. Mainster, and J.S. Parker: Vision Res. 20, 1099 (1980)
2   R.H. Webb, G.W. Hughes: IEEE Trans. Biomed. Eng. BME-28, 488 (1981)
3   R.H. Webb, G.W. Hughes, O. Pomerantzeff: Appl. Optics 19, 2991 (1980)
4   M.A. Mainster, G.T. Timberlake, R.H. Webb, and G.W. Hughes: Ophthalmology
    89, 852 (1982)
5   G.T. Timberlake, M.A. Mainster, R.H. Webb, G.W. Hughes, and C.L. Trempe:
    Invest. Ophthalmol. Vis. Sci. 22, 91 (1982)

Quantification of the Shape and Color of the Optic Nerve Head

Tom N. Cornsweet, S. Hersh, J.C. Humphries, R.J. Beesmer, and
D.W. Cornsweet

University of California, Irvine and Rodenstock Instruments Corporation
Laguna Hills, CA 92653, USA

Introduction

Many pathological conditions of the eye and of the body as a whole cause
changes in the color and three-dimensional shape of the ocular fundus. Glaucoma is probably the best known of these conditions. (For some examples,
see [1] through [9].) It causes increases in the pallor and cupping of the
optic nerve head. Therefore, it would obviously be of great potential value
for research on and diagnosis and treatment of these diseases if the shape
and color of the fundus could be measured with precision, and several different procedures have been developed at various laboratories around the
world to try to do that ([10] through [34]).

We have been wrestling with the problem, or rather with the many problems
for a long time now, and have finally progressed far enough to begin to obtain useful results. I will first describe the basic elements of the instrument we are developing, and then present some results for the quantification
of pallor and cupping of the optic disk.

Methods

The basic elements of the instrument are shown in Fig. 1. We have combined
some of the techniques that have been described in the literature and have
tried to correct some of the deficiencies of earlier techniques, while generating some new deficiencies of our own. To measure color, we illuminate
the fundus with light of wavelengths that are selectable by filters, as many
other people have done. But instead of using a standard fundus camera and
photography, we have designed a special optical system, and the pictures are
taken by a sensitive television camera. The pictures are then processed by
a microcomputer.

For the measurement of color, our special optical system has three major
advantages over the more standard fundus photography techniques. First, we
are able to introduce into our pictures what we call a reference spot that
enables us to measure and compensate for any variations in the transmissivity
of the optical system and sensitivity of the television camera that can occur
from time to time and from color to color. Second, the optical system automatically tracks the eye and keeps the optical axis of the instrument centered on the patient's pupil to ensure that the light is not vignetted by
the pupil. Third, the television camera we use has a response that is linear
with light intensity, so that we do not have to contend with the strong non-
linearity and variability of photographic film. (The television camera is
also far more sensitive than film.)

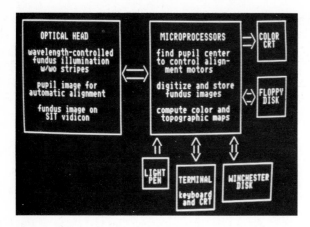

Fig. 1. Basic elements of the instrument and their interconnections

To measure depth, we project a set of bright stripes on to the fundus and obtain a simultaneous stereo pair of images on the television camera. The microcomputer then analyzes the stereo pairs and generates depth profiles and contour maps from them. Here, our special optical system has other advantages over more standard systems. First, the fact that the instrument axis is always centered on the pupil removes an important potential source of variability, both because it avoids vignetting and because the variations in refractive power that occur across the normal entrance pupil [35,36] do not cause variations in the results. Second, our optical system, like all optical systems, produces some geometrical distortion of the fundus pictures, and the television camera also distorts the pictures. These distortions, in turn, would cause substantial distortions of the calculated depth values. Therefore, we have built into our instrument a system that permits us to take special pictures at any time from which the microcomputer can calibrate and remove the effects of distortion.

Pallor Measures

Figures 2 and 3 are photographs of the television monitor displaying the fundus pictures. Fig. 2 was taken with green light and Fig. 3 with red light. The features of the fundus that appear red ophthalmoscopically, like the vessels, are those that reflect red light and absorb green light, so they appear light in both pictures. In the picture with red illumination, the contrast is low because both the red vessels and the whiter background reflect more or less the same amount of red light, and in the other picture, the background, being whiter, reflects more green light than the vessels.

To quantify pallor, we define the pallor at each point in the picture as the ratio of the green reflectance to the sum of the red plus green reflectances:

$$\text{Pallor} = P = \frac{\text{Green Reflectance}}{\text{Red Reflectance} + \text{Green Reflectance}} \; .$$

By this definition, a pure red spot would have a value of zero, pure white one-half, and pure green one.

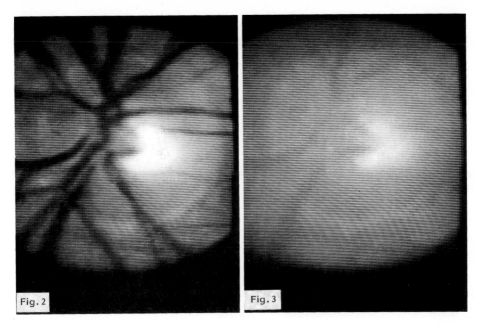

Fig. 2. Photograph of the television monitor while the fundus is illuminated with green light

Fig. 3. The fundus under red illumination. The contrast among fundus features is low at these wavelengths

Figure 4 is what we call a pallor picture. It is generated in three steps. First we slide the red picture over the green one to register them, in order to correct for any eye movement that might have occurred between the times they were taken. Then the computer measures the reference spot on each picture and corrects the pictures against differences in sensitivity. Finally the computer calculates a new picture--the one in this figure is an example-- such that the intensity at each point in the new picture is proportional to the P value in the formula above.

Now this pallor picture actually contains a quantitative measure of pallor at each imaged point on the fundus, but it isn't very informative just to look at. We can read off the value of P at any point in the picture by touching that point on the TV screen with a light pen, but a more convenient way of visualizing the distribution and extent of pallor is to color code the picture, as shown in Fig. 5. This picture is a color-coded version of Fig. 4. What the computer has done is to make all the P values between zero and .05 dark red, all the values between .05 and .1 medium red, between .1 and .15 orange, etc. In other words, different colors in the picture represent different ranges of pallor. We have come to call pictures like this pallor chromographs.

We have measured the repeatability of the pallor pictures by taking a series of them on the same eye and then measuring the variability of the P values at a lot of different points chosen at random. The standard deviation of the values is 2.8% of the range from pure red to pure white. Therefore, a change in pallor of about five and a half percent would be detectable 95% of the time. We are now working on procedures to reduce this variability further.

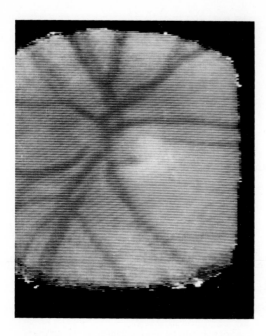

Fig.4. A "Pallor picture". The
brightness at each point is
proportional to the value of the
quantity "Pallor", as defined
in the text

▼ Fig.5. The pallor picture in
Fig.4 color coded to display,
quantitatively, the distribution
of pallor

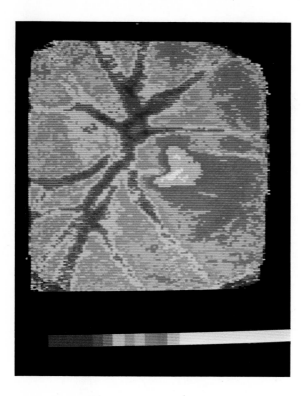

Depth Measures

Figure 6 shows a stereo pair of images of the fundus on which green stripes have been projected. The microcomputer analyzes the picture to produce a set of depth profiles, as shown in Fig. 7. The depth profiles on the right are sections taken through the regions of the fundus shown by the straight vertical lines on the left. The white horizontal line on the left represents a depth of about 500 microns.

We have measured the repeatability of this procedure by taking a number of separate pictures and comparing the resulting profiles. Five profiles taken independently on the same region of the disk are superimposed in Fig. 8. The repeatability does vary somewhat from one patient to another and from one place to another on the same fundus.

There are two sources of variability that we have identified. First, wherever there are highlights, especially from the veins, the computed depth value is not really the depth of the vessel surface and can vary strongly with pulse. Therefore, the depth values at the vessels are suspect.

A second reason for variability is more interesting. There are regions of the optic disk that contain no resolvable features, even in the best of fundus photographs. In those regions, a stereo pair cannot contain any depth information. The reason that we project stripes onto the fundus is to provide what in effect are features in those otherwise featureless regions, so that depth information can be extracted from the pictures. However, in some of those regions, the projected stripes are almost invisible. One of those regions is marked by the arrow in Fig. 6. In those regions, our depth profiles become strongly variable.

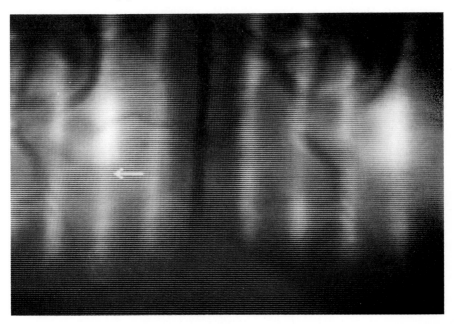

Fig. 6. Simultaneous stereo pair of images of a fundus on which green stripes have been projected. The arrow points to a region in which the tissue reduces the contrast of the stripes

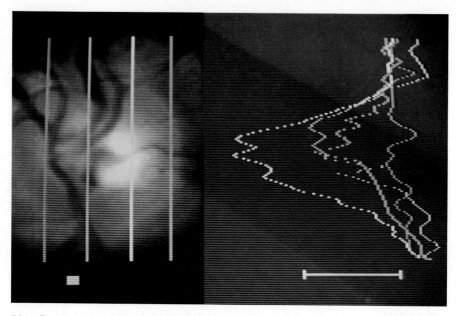

Fig. 7. Depth profiles computed from the picture in Figure 6. The profiles on the right are sections through the regions that are indicated by vertical lines of corresponding colors on the left. The horizontal line near the bottom indicates a depth of about 0.50 mm

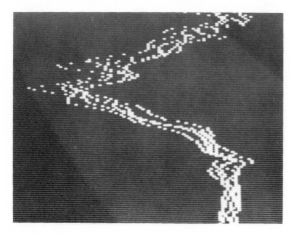

Fig. 8. Repeatability of the depth profiles. Five independently measured profiles through the same region of the optic disk are superimposed

When we select points on the profiles that do not lie on vessels or on regions where the stripes disappear, the standard deviation of the depth measures is 5% of the range of depths we see in a typical normal fundus. Therefore, a change in depth of 10% at any point would be detected 95% of the time. When we include all points, regardless of vessels and the disappearance of stripes, the standard deviation increases to about 8%.

We mentioned above that the optical and electro-optical distortions in the raw pictures are corrected. To illustrate how important these corrections are in our system, Fig. 9 shows our correction profiles to the same scale as the fundus profiles in the earlier slides. These profiles can be interpreted as the depth profiles that our instrument would produce from a stereo pair of pictures of a perfectly flat fundus. (More accurately, the latter are the profiles that would result from a fundus that lay entirely in the focal plane of the eye.) The curvature of each profile and the horizontal displacements between some of them are produced by distortion in our optics and television camera, and would distort the fundus profiles that much if they were not corrected. Despite the fact that this distortion looks very strong, our system is not unusually badly distorted. Actually, it has very little distortion. It is just that all stereometric procedures for fundus measurement are extraordinarily sensitive to distortion. The maximum distortion represented in this slide is actually caused by a geometric distortion of only 1.5%.

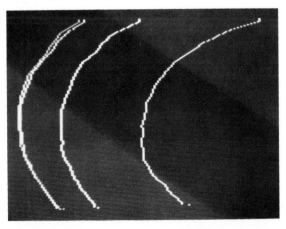

Fig. 9. A plot of the distortion introduced by the camera and optical system of the instrument. If this distortion were not compensated, it would result in serious errors in the computed fundus depth profiles

Figure 10 shows a contour map of the same optic disk that is shown in Figs. 6 and 7. It was generated in five steps. First, a stereo pair was stored when the disk was illuminated with stripes. Second, the stripes were moved sideways a little on the fundus, so that they occupied places between the locations of the stripes on the first picture, and another stereo pair was stored. Third, the two sets of pictures were registered to correct for any eye movement that might have occurred between them. Fourth, the computer generated profiles for both pictures and corrected them against distortion,

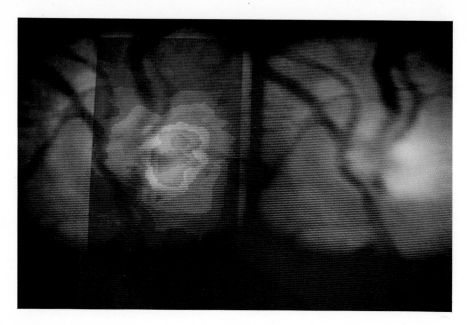

Fig. 10. A color coded contour map of the optic disk. The different colors represent 100 micrometer depth intervals

and finally, the computer combined the two sets of profiles and computed and colored the contours shown in this figure.

Summary

We have developed an instrument that produces measures of the shape and color of the fundus. The repeatability of the results implies that the instrument has promise in research and in the detection of glaucoma and other diseases that affect the eye.

References

1   M.F. Armaly: Invest. Ophthalmol. 9, 425 (1970)
2   I. Goldberg: Aust. J. Ophthalmol. 9, 223 (1981)
3   K. Iwata: Docum. Ophthalmol. Proc. Series 19, 233 (1981)
4   J.E. Pederson and D.R. Anderson: Arch. Ophthalmol. 98, 490 (1980)
5   H.A. Quigley and W.R. Green: Ophthalmology 86, 1803 (1979)
6   B. Schwartz: Arch. Ophthalmol. 89, 269 (1973)
7   A. Sommer, I. Pollack, and A.E. Maumenee: Arch. Ophthalmol. 97, 1444 (1979)
8   G.L. Spaeth, R.A. Hitchings, and E. Sivalingam: Trans. Am. Acad. Ophthalmol. Otolaryngol. 81, OP217 (1976)
9   M.E. Yablonski, T.J. Zimmerman, M.A. Kass, and B. Becker: Am. J. Ophthalmol. 89, 585 (1980)
10 H. Alanko, E. Jaanio, P.J. Airaksinen, and H. Nieminen: Acta Ophthalmol. 58, 14 (1980)
11 B. Bengtsson and C.E.T. Krakau: Acta Ophthalmol. 57, 503 (1979)
12 P. Betz, F. Camps, C. Collignon-Brach, and R. Weekers: J. Fr. Ophthalmol. 4, 193 (1981)

13 J.L. Calkins and C.D. Leonard: Invest. Ophthalmol. 9, 458 (1970)
14 J.S. Cohen, R.D. Stone, J. Hetherington, Jr., and J. Bullock: Am. J. Ophthalmol. 82, 24 (1976)
15 T.J. Ffychte, A.R. Elkington and I.J. Dowman: Trans. Ophthalmol. Soc. U.K. 93, 251 (1973)
16 J. Gloster and D.G. Parry: Brit. J. Ophthalmol. 58, 850 (1974)
17 H. Goldmann and W. Lotmar: Klin. Mbl. Augenheilk. 176, 547 (1980)
18 O.C. Holm, B. Becker, C.F. Asseff, and S.M. Podos: Am. J. Ophthalmol. 73, 876 (1972)
19 E. Jaanio, H. Alanko, P.J. Airaksinen, H. Nieminen, and S. Lahde: Acta Ophthalmol. 58, 7 (1980)
20 C.A. Johnson, J.L. Keltner, M.A. Krohn, and G.L. Portney: Invest. Ophthalmol. Vis. Sci. 18, 1252 (1979)
21 C.-H. Jonsas: Acta Ophthalmol., Suppl. 117, 9 (1972)
22 M.S. Kottler, A.R. Rosenthal, and D.G. Falconer: Invest. Ophthalmol. 96, 116 (1974)
23 C.E.T. Krakau and K. Torlegard: Acta Ophthalmol. 50, 863 (1972)
24 M.A. Krohn, J.L. Keltner, and C.A. Johnson: Am. J. Ophthalmol. 88, 859 (1979)
25 G.L. Portney: Trans. Am. Acad. Ophthalmol. Otolaryngol. 81, OP231 (1976)
26 R. Rochels: Ophthalmologica 181, 139 (1980)
27 M. Mikuni, H. Yaoeda, S. Fujii, and M. Togano: Acta Med. Biol. 19, 204 (1972)
28 R. Rochels: Ophthalmologica 180, 277 (1980)
29 A.R. Rosenthal, D.G. Falconer, and P. Barrett: Arch. Ophthalmol. 98, 2027 (1980)
30 K.E. Schirmer: Klin. Mbl. Augenheilk. 164, 688 (1974)
31 B. Schwartz: Trans. Am. Acad. Ophthalmol. Otolaryngol. 81, OP227 (1976)
32 B. Schwartz, N.M. Reinstein, and D.M. Lieberman: Arch. Ophthalmol. 89, 278 (1973)
33 W. Weder: Klin. Mbl. Augenheilk. 176, 165 (1980)
34 R.L. Wiggins, K.D. Vaughan, and G.B. Friedmann: Arch. Ophthalmol. 88, 75 (1972)
35 A. Ivanoff: J. Opt. Soc. Am. 46, 901 (1956)
36 G. van den Brink: Vision Res. 2, 233 (1962)

Part 5

Developments in Visual Electrodiagnostic Techniques

Electrical Soundings in the Visual Current

Irwin M. Siegel

Department of Ophthalmology, New York University Medical Center
New York, NY 10016, USA

"Light breaks where no sun shines. . ." Dylan Thomas

The reconstruction of visual space within the cortex of the brain is the culmination of a series of complex events of which we possess only the most rudimentary pieces. We can identify with certainty the initiating photochemical processes which result from quantal absorption within the rods and cones but we are less certain of the chemistry responsible for converting that absorption into a receptor potential which is the source of the electrical current. While microelectrode studies have provided excellent intracellular records from each neural element as the flow of current proceeds to the optic nerve fiber, remote recording from the surfaces of the eye can only detect the confused echos of the original signals from each layer. Further along the stream of visual activity, electrical signals (now transmitted as spikes) must accurately define extent, size, orientation, contrast, and color of the original stimulus. The subtle interaction within and between receptive fields in LGN and cortex requires an order of neural processing that we may only guess at. Finally, we must try to conjure up the "visualization" of external objects at the conscious level, a process that must involve the fabrication of an electrical web incorporating all the information necessary to identify and locate the object in space.

In view of the complexity of the electrical events associated with perceiving even the simplest objects, it would be outrageously optimistic to conceive of simple clinical tests using remote electrodes that would somehow be quantitative in nature, assist in localizing the disease process and provide definitive diagnostic assistance. Yet, over the past 40 years such techniques were developed and are now considered routine tools in retinal studies. The fundamental ideas incorporated in the techniques were conceived in very distinct steps, and most of them were taken by early workers after careful consideration of the highly organized nature of the retina itself. For example, the distinct properties of rods and cones provided the information required to vary appropriately the color and frequency of the stimulus and the adaptation state of the retina to distinguish electrically between the two types of photoreceptors. But of even greater concern to the early workers was the nature of the waveform itself, for they intuited that if the individual components of the ERG could be identified, it would greatly enhance its diagnostic application.

It was GRANIT [1] who demonstrated that the complex electrical changes evoked from the retina in response to light derived from specific assemblies of cells. Using a technique designed to produce selective anoxia of the retinal layers he discovered, among other things, that the deep, negative potential originated from the photoreceptors and the positive b wave, from the mid-retinal layers. The simplicity of this arrangement allowed vision researchers quite early on to utilize the electroretinogram (ERG) as a diagnostic tool for studying several different kinds of ocular diseases [2]. The development of electrical tests of visual function for the last 50 years provides similar revealing examples of interactions between psychophysics and physiology, between clinician and laboratory scientist.

Early ERG Recording

Interestingly, it was not until the introduction of plastic contact lenses in the 1940's that routine human ERG recordings became feasible [3]. Contact lenses of either the full scleral type, or the later design incorporating a lid speculum [4] provided a stable, comfortable platform, for a small piece of silver or stainless steel to make electrical contact with the precorneal tear film.

The clearly divisible behavior of the rod-cone retina into dark and light adapted function was an obvious starting point for obtaining ERG records. It was also clear to the early researchers that stimulus conditions could be arranged so as to excite selectively the rod or cone system. So, for example, a bright light, flickering at 20 times or more a second would certainly evoke a flicker-following response from the cone system. Likewise, a dim, blue flash can be shown to produce a predominantly rod ERG response. Manipulating stimulus color and the state of retinal adaptation led to further, unambiguous, separation of the two classes of retinal photoreceptors. One should bear in mind, however, that these responses represent activity from a large population of cells from virtually the entirety of the retina. Therefore, while it is quite simple to separate rod and cone activity, it is not possible to assign the recorded activity to a particular geographic location. Cone activity from the macula, for example, could not be distinguished from cones located outside that area. In diseases which grossly affect either the rod or cone system, however, the fact that the ERG is so diffuse becomes a decided advantage. Visual loss due to a generalized cone degeneration results in an absent flicker ERG, while an equivalent loss due to an isolated foveal lesion hardly affects the recorded response [5]. Similarly, the waveform characteristics of the ERG helps distinguish a variety of nightblinding diseases and retinal vascular disorders [6].

Origin of ERG Potentials

It is fair to say that the diagnostic power of ERG recording is fully realized when it is known from which portion of the retina each constituent portion of the waveform derives. Figure 1 shows that the initial, negative-going a wave (actually the leading edge of the larger receptor potential) derives, as originally demonstrated by GRANIT [1], from photoreceptor activity, while the larger, positive, b wave has its origins from activity of cells in mid-retinal layers. There has been evidence [7] that the recorded b-wave potential results from glial cells responding to changes in potassium concentration occurring in and around the bipolar cell region. It is of interest to note that even though the b wave does not seem to derive directly from a cell in the mainstream of excitation, it still remains a sensitive, reliable, albeit an indirect reflection of retinal function. Moreover, the dependence of the mid-retinal layers (and the b wave) on the retinal circulation and the similar dependency of the photoreceptors on the choroidal supply makes for a rapid diagnostic tool for determining the integrity of either vascular system. For instance, if the central retinal artery is occluded even to a minor extent, the b wave suffers an immediate and obvious decrease in amplitude [6]. On the other hand, if the choroidal circulation is occluded, a more profound change in the ERG occurs, since the source of

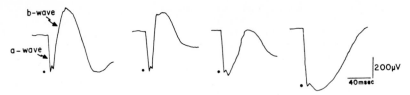

b-wave

a-wave

200μV

40msec

Fig. 1 Unmasking of the complete a wave and photoreceptor potential by inducing a state of retinal anoxia which slowly abolishes the b wave.

all visual excitation, the photoreceptors, as well as more proximal layers are directly affected. In the latter circumstance, both a- and b-wave formation would be directly and severely affected [8].

The distinct separation of activity from mid and outer retinal layers in certain types of vascular disease suggested that the sorts of waveform changes accompanying other types of pathology might also be helpful in identifying the lesion site within the retina. The first clearcut demonstration that this was a feasible approach occurred in patients with congenital stationary nightblindness (CSNB). Using direct fundus reflectometric measurements of rhodopsin, ERG and electrooculographic (EOG) studies, and psychophysical tests of retinal adaptation, several forms of CSNB were examined over a 12-year period [9,10,11]. One particular form of CSNB, transmitted as a recessive trait, and producing no fundus abnormality will serve as an example of the general approach.

Figure 2 shows the results of some electrical testing and also the dark adaptation curve obtained from a patient with this more common form of CSNB. The dark adaptation curve confirms in a quantitative way the main complaint

CSNB with Normal Fundus

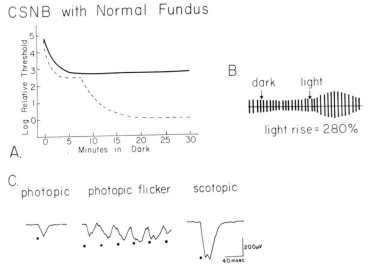

B.

dark   light

light rise = 280%

A.

C.

photopic   photopic flicker   scotopic

200μV

40msec

Fig. 2 Dark adaptometry and electrical findings of a patient with congenital stationary nightblindness with no fundus changes.

of the patient, i.e., he can see well in the light but very poorly in the dark. It is important to note that unlike most generalized degenerations of the retina which produce marked elevations in both rod and cone thresholds, the cone sensitivity of the CSNB patient in Fig. 2 is normal. The ERG shows a predominantly negative response with little sign of a b wave in the light or dark. The EOG in this patient is normal which alone would indicate that this is neither a generalized degeneration nor a central retinal artery occlusion [6]. Not shown in the figure is the fundus reflectometry which in this patient showed a normal density of rhodopsin in the rods as well as a normal capacity to regenerate after being bleached away. Furthermore, the normal amplitude a wave (which is the leading edge of the photoreceptor potential) indicated that the rod structure was also intact. It was concluded from all these results that the heritable defect responsible for CSNB did not involve the rod receptors or the visual pigment rhodopsin. The loss of b-wave activity strongly suggested that the lesion was some sort of transmission defect (a neurochemical abnormality) located in the mid-retinal layers. Thus, the CSNB studies laid the framework for a limited layer-by-layer analysis based on functional dissection of retinal activity.

In the last 20 years a great deal of refinement in identifying the electrical potentials arising from virtually each cell in the retina took place [12]. But interestingly, the continuing evolution of the layer-by-layer concept came from two quite disparate test procedures: cortical responses evoked by patterns of light and attempts to evaluate objectively electrical activity from local retinal areas.

Cortical Recordings

While recordings of visual evoked potentials (VEP) had been performed for many years, it was not until the stimuli were made sufficiently interesting to the brain that VEP testing developed as an accurate diagnostic procedure, one which correlated rather well with subjective acuity measures [see ref. 13 for a review]. The scalp recording site is over area 17 of the occipital cortex. This particular location, of course, is the terminus for light activated electrical signals originating in the retina. Foveal representation is so great in area 17 compared to other retinal areas that the VEP is heavily dominated by activity arising in the small foveal region. This is true (to a lesser extent) even if a diffuse light is presented to the eye. However, it was the development of phase-reversing patterns of stripes or checkerboards which allowed the clinician to ascertain foveal function without interference from activity evoked from surrounding retina [14]. Reversal patterns work on the principle that only portions of the retina underlying the imaged stripes or checks are stimulated as the patterns alternately shift a half cycle (180° phase shift) to the left and right. Since the same number of light and dark areas are present before and after the pattern reversal, the overall screen luminance does not change. Therefore the scattered light from the reversing pattern is perceived by the outlying portion of the retina as steady and does not interfere with the development of electrical potentials developed by retinal areas underlying the alternating pattern.

The VEP is widely used as a particularly accurate measure of eye-to-cortex conduction time. Even small latency increases in the time-to-peak of the VEP waveform have been shown to have diagnostic relevance for detecting diseases which affect the myelin sheath of the optic nerve fibers, as well as other conditions associated with visual loss [15]. The electronically generated patterns on the TV screen that are used for stimulating the VEP are so versatile, that many parameters may be explored. To obtain a rough approximation of Snellen acuity, for example, one may choose to decrease the size of a high contrast checkerboard pattern to the point of response extinction. An even more satisfying way of objectively defining visual resolution is working with a moderately small (30 min) check-size and then decreasing contrast. The latter technique is also an extremely sensitive approach to measuring subtle latency changes. The patient in Fig. 3 has mild one-sided retrobulbar neuritis. Starting at the top, the pairs of tracings show the VEP responses of the right and left eye to reversing checks of lower and lower contrast. At high contrast levels (upper tracings) a small increase of latency can be seen in the left eye as compared with the right, but the amplitudes for both are about the same. As the contrast of the pattern is decreased much greater latency differences are revealed in the mildly affected eye as well.

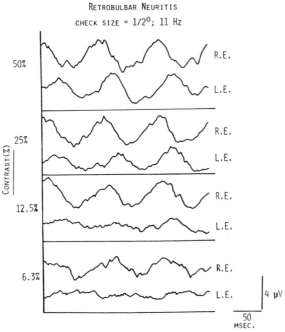

RETROBULBAR NEURITIS

CHECK SIZE = $1/2^\circ$; 11 Hz

Fig. 3 VEP amplitudes as a function of pattern contrast in a patient with unilateral retrobulbar neuritis.

Considering all the studies using the VEP as an objective measure of acuity, contrast sensitivity, and so on, one might surmise that what is being recorded is the massed discharge of individual cortical cells in area 17 responding to retinal stimulation. There is a large body of data which indicates otherwise. The classic experiment was performed by KELLY and VAN ESSEN [16]. When a cortical cell was stimulated and issued appropriate spike discharges, nearby groups of glial cells that responded to the K^+ changes and produced a field change that most resembled the VEP records obtained from scalp recording. Thus it would appear that the VEP, like the ERG b wave [7] is a reflection of glial cell activity, one step removed from the primary site of neural activity. Quite by chance, we discovered supporting evidence in humans with optic nerve glioma. The latter retained normal or near normal visual acuity but had no recordable VEPs [17]. This rather startling finding is exactly the opposite sort of situation BODIS-WOLLNER et al. [18] reported in their patient who was blind but retained a normal VEP! Their data is easily explained on the basis of a higher center lesion, and a CT scan revealed that a tumor had in fact destroyed parts of cortical areas 18 and 19. So, visual processing up to area 17 was intact in their patient but those areas of the brain concerned with visual recognition were damaged and prevented the imagery from emerging to consciousness. I believe that in our patients with gliomatous tissue changes, the primary cortical cells in area 17 are intact but the glial cells (which form a syncytial type structure in the brain [19]) produce VEP potentials that can no longer generate the K^+ — dependent signals normally recorded by the remote electrodes on the scalp.

Pattern Stimulation and ERG Recording

Reversal pattern stimuli were actually developed by RIGGS and his co-workers [14] for recording from a localized retinal area of retina. In clinical studies, pattern ERG (PERG) amplitudes seemed to decrease markedly if the macula was impaired in some way [20]. The finding appeared to support the assumption that decreased macular function results in decreased electrical activity. However, it quickly became apparent that the pattern ERG had some properties one never associated with the ERGs evoked by diffuse light stimuli. For example, it was found that refractive error and target contrast were important variables [21]. In fact, it was found necessary to use active eye electrodes that would in no way interfere with the optics of the eye. Indeed, most laboratories nowadays use a goldfoil electrode hooked over the lower lid [22], or an electrically conducting thread placed in the lower fornix or draped around the cornea in the limbal region [23]. Not only does defocusing decrease the PERG amplitude, but many forms of amblyopia also result in PERG reduction [24]. In an attempt to understand better the origin of the PERG potential, Maffei and Fiorentini performed an animal experiment in which they sectioned one optic nerve and followed the ERG response of the eye to patterned and diffuse light stimuli for a period of some months [25]. After a few weeks, the PERG amplitudes diminished and after a few months were unrecordable. Electrical responses to diffuse flash stimuli during this period remained unchanged. Such data would lead one to conclude that ERG response to patterns derives at least in part from the inner retinal layers. If this were true then patients with optic nerve neuropathy, including those with glaucoma, would show reduced PERG activity. Several reports [26,27,28] indicate that this is certainly the case.

In using the PERG for diagnostic purposes it is important to be aware of the fact that the majority of ganglion cells are located in the 15-20 degree central area of the retina (see Fig. 4). The ganglion cell layer is several layers thick in and around the fovea, then thins out to a single layer containing far fewer cells outside the macular region. If the PERG amplitude is related (even in a crude way) to numbers of ganglion cells stimulated then the response should be "center weighted." Therefore, if the center of the pattern was occluded one should obtain a greater loss in amplitude compared to the response elicited by a pattern with an eccentrically located, but equal-sized, occluded area. Figure 5 shows this effect and also reveals that the PERG center weighting effect is not as great as that seen for the VEP. But this is hardly surprising given the great foveal representation in area 17 and the proximity of this portion of the cortex to the recording electrodes on the overlying scalp.

Briefly stated then, the retina seems capable of generating responses attributable to the outer or inner layers depending on whether the stimulus consists of a diffuse flash or local luminance changes of the sort associated with a reversing pattern display.

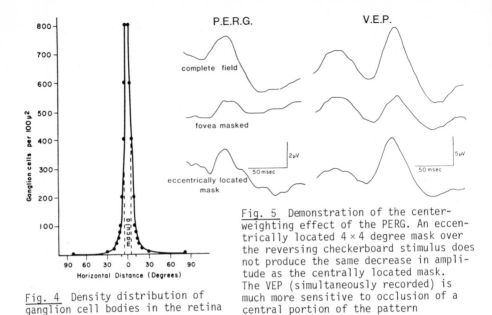

Fig. 4 Density distribution of ganglion cell bodies in the retina

Fig. 5 Demonstration of the center-weighting effect of the PERG. An eccentrically located 4 × 4 degree mask over the reversing checkerboard stimulus does not produce the same decrease in amplitude as the centrally located mask. The VEP (simultaneously recorded) is much more sensitive to occlusion of a central portion of the pattern

Focal ERG Recording from the Macula

If the PERG potentials derive mainly from the ganglion cell population located in the macular region, how can they be separated (in the normal eye) from electrical activity arising solely from the outer cell layers of the same area? The answer seems to be by recording focal ERGs confined to the macular areas using stimuli which undergo purely luminance changes. There is a long history involving many researchers concerning the recording of focal electroretinograms (FERG). However, in all these studies there is one overriding consideration--to prevent retinal areas outside the stimulus area from contributing to the response. Typically, the stimulus was surrounded by a steady light background. The latter maneuver served to adapt and hence desensitize adjacent retinal elements to the effects of stray light scattered by the stimulus. To test the efficacy of the surround, the stimulus could be arranged to fall on the optic nerve head [29]. If no response was recorded it was assumed that the steady surround was of sufficient brightness to prevent stray light from producing detectable retinal potentials. The experimental arrangement makes it ideal for recording photopic responses, particularly from the cone-rich central areas but next to impossible for eliciting local rod responses in the retinal periphery.

One particular approach, used by AIBA et al. [30] for obtaining the FERG provided us with the starting point for a new design. Aiba and his co-workers used a glow modulator tube driven electronically to produce a sinusoidally modulated light source. The light output was transmitted through a red filter and viewed against a large steady surround. We have constructed a source consisting of several dozen red LEDs packed in a circular array and placed in a ganzfeld bowl. The LEDs are driven by a waveform generator via an op-amp current driver. In this way, we provide a monochromatic light source of known wave length which can be varied in wave shape, modulation depth, and temporal frequency. In addition, the d-c level of the LEDs can be changed

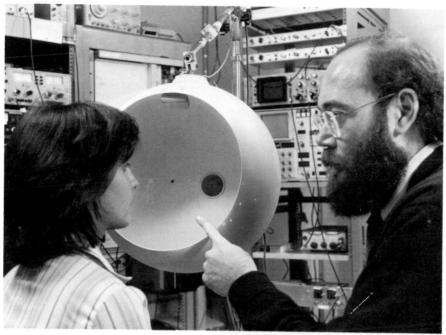

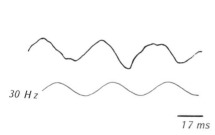

30 Hz

17 ms

Fig. 6 Apparatus for producing the focal ERG. An LED array is located in the center of a ganzfeld bowl. The upper trace shows the FERG in response to a 30 Hz sinusoidally driven stimulus. Inset at lower right shows the gold foil electrode hooked over the lower lid of the subject.

so that the sinusoidal modulation can be varied around several mean brightness values. The apparatus with some sample FERG tracings are shown in Fig. 6. As the LED display executes a waxing and waning (shown in the bottom trace of each pair), the FERG (upper trace of each pair) follows along as a slightly out-of-step sinusoidal response. It is a simple matter to compare the stimulus and steady-state response and arrive at a very accurate measure of latency (or more rigorously) the phase difference between the two. Since changes in the modulation depth result in amplitude decrements, it is possible to define fully the dynamic responsivity of the FERG by varying modulation over a large range of temporal frequencies. We have utilized the remarkable versatility of the LED stimulus array to derive modulation transfer functions (so-called deLange curves) psychophysically and compare them with the similarly derived FERG function on the same apparatus. For purposes of this presentation, however, it is only necessary to emphasize that the luminance-varying display elicits a focal ERG from the macula which is quite distinct in character from the pattern-evoked ERG. The FERG can be thought of as a small potential elicited from the outer and mid-retinal layers while the pattern ERG is a reflection of ganglion cell activity stimulated by brisk changes at the borders of a highly textured stimulus. It is also fair to say that the PERG is not a direct representation of ganglion cell firing, i.e., of the spikes produced by the cells themselves. Most likely, the PERG, like most retinal (and cortical!) activity recorded from remotely located electrodes is a reflection of light-induced glial cell activity.

A Layer by Layer Electrical Analysis of Visual Function

To understand fully visual loss in a patient, there are three pieces of information which the clinician must obtain. First, the extent of the loss must be quantified--whether it be decrease in acuity, field changes, or color abnormality. Second, the location of the disease must be ascertained. If routine observation indicates changes within the retina, then we may ask which layer is affected, or how much of the retina is affected. Rarely does ophthalmoscopic or even sophisticated photographic evaluation give a precise correlation of dysfunction. If we suspect that the disorder affects postretinal structures, we may ask whether the optic nerve is directly or indirectly affected. Or perhaps the visual system up to area 17 is competent but that cortical processing of visual information is ineffective. Third, the clinician would seek to determine, after quantifying and localizing visual loss, the nature of the pathology. Of what category is the underlying pathology--vascular, abiotrophic, iatrogenic? Or is a quirk of genetic fabrication at fault? While the latter quest may be the ultimate goal of diagnostic procedures, defining the location and extent of the lesion presents us with the immediate challenge.

By way of defining some basic concepts, consider the flow of information from eye to brain shown in Fig. 7. Since the excitation begins at the eye

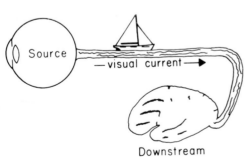

Fig. 7 The visual current starts in the photoreceptors and flows downstream toward the brain

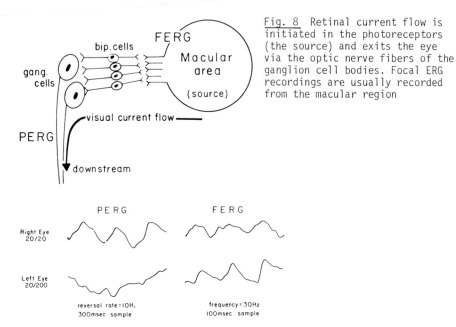

Fig. 8 Retinal current flow is initiated in the photoreceptors (the source) and exits the eye via the optic nerve fibers of the ganglion cell bodies. Focal ERG recordings are usually recorded from the macular region

Fig. 9 Recording of the PERG and the FERG from a patient with a one-sided retrobulbar neuritis. The FERG is normal in each eye while the PERG is absent in the affected eye.

I define the retina, and more specifically, the retinal photoreceptor, as the current source for the flow. The current flows like any stream towards a sink, which in the case of vision is area 17 of the visual cortex. Working backwards, area 17 is downstream of the LGN which is downstream of the optic nerve and ganglion cell bodies. Figure 8 shows a detailed schematic of the current flow in the retina, emphasizing a particular region, the macula. If the macula is nonfunctional or severely affected, no downstream current flows--and visual activity further along is nil. The FERG of course is absent. The differences between the FERG and PERG, along with the specific locations suggested by these differences, are outlined in Table 1. If, as suggested in the table, a disease affects only the inner retinal layers, the FERG should be present and the PERG absent. Figure 9 shows the two potentials recorded from the eyes of a patient with a one-sided retrobulbar

Table 1 Separation of the origin and clinical applications for the focal and pattern ERGS

	ORIGIN	APPLICATION
FOCAL ERG (represents source activity from macula)	Photoreceptor and Mid-retinal layer activity	Abnormal if lesion is in outer retinal layers.
PATTERN ERG (downstream reflection of source activity)	Ganglion Cell layer activity	Abnormal if lesion affects inner retinal layers, directly or indirectly.

161

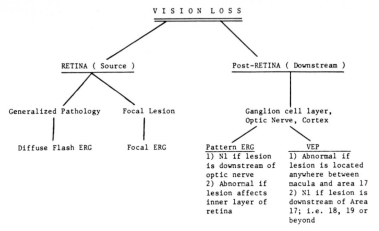

VISION LOSS

RETINA (Source)   Post-RETINA (Downstream)

Generalized Pathology   Focal Lesion   Ganglion cell layer,
                                                Optic Nerve, Cortex

Diffuse Flash ERG   Focal ERG   Pattern ERG   VEP
                                          1) Nl if lesion   1) Abnormal if
                                          is downstream of   lesion is located
                                          optic nerve   anywhere between
                                          2) Abnormal if   macula and area 17
                                          lesion affects   2) Nl if lesion is
                                          inner layer of   downstream of Area
                                          retina   17; i.e. 18, 19 or
                                                                   beyond

Fig. 10   Flow chart showing how the clinician may proceed with testing procedures on patients with visual loss.

neuritis. Note that in the affected left eye, the ERG in response to a pattern is absent while the focal luminance ERG amplitude is about the same as
the unaffected eye. In this patient, a clear separation between the normal
outer and mid-retinal layers and the affected inner layers can be made. If
the photoreceptors of the macula were disturbed, downstream current to the
ganglion cells located in the inner retina will be markedly attenuated or
absent.

     Figure 10 shows, in flow chart fashion, the manner in which one may pro-
ceed to localize loss of acuity in the visual system. Ophthalmoscopic findings give the first clue as to whether the problem appears to be in the retina or located downstream of the retina. If retinal changes are consistent
with a generalized retinopathy, a diffuse flash (ganzfeld) ERG will show
appropriate amplitude decrements using scotopic and photopic stimulus conditions. However, it is when the diffuse flash ERG is normal in a patient
with vision loss that the search for a diagnosis becomes interesting. A
normal response to a diffuse flash ERG may occur even in patients with obvious foveal lesions, since the actual number of cones in the fovea represents a very small contribution to the total response. If the fovea appears
disturbed by ophthalmoscopic observation or fluorescein angiography, a focal
ERG will provide a confirmatory estimate of foveal function and indicate
that the lesion level is in the outermost of the retinal layers. A more
interesting situation arises when there is visual loss, the fundus appears
normal, and the diffuse and focal ERG records show normal waveforms. Since
this state of affairs suggests a post-retinal condition, we follow the right-
hand set of branchings in the flow chart of Fig. 10. The key test is the
pattern ERG. An abnormal PERG signifies a loss of inner retinal activity,
e.g., a primary optic nerve disease [25], a compressive lesion of the optic
nerve [27], or glaucoma [28]. There is no point testing further downstream,
that is, at the cortical level using the VEP because the visual current is
already compromised upstream. If a VEP record was obtained in the absence
of a PERG it would of course be attenuated--but this abnormal response could
not be ascribed to a cortical defect. The decreased VEP amplitude, except
in rare instances of cortical tissue damage, simply confirms a loss of central vision. Such loss could be associated with a disease affecting the
photoreceptors (at the "source") or anywhere downstream past that level along

Table 2 Various types of disease entities associated with visual loss and how the various test modalities aid in diagnosis and lesion location.

Condition Producing Vision Loss	Ganzfeld ERG	Focal ERG	Pattern ERG	VEP
Generalized rod/cone degeneration with the macula involved	Absent	Variable	Decreased	Decreased
Macular degeneration	Normal	Decreased	Decreased	Decreased
Generalized cone degeneration with macula spared	Photopic ERG absent	Variable	Variable	Variable
Optic nerve neuropathy	Normal	Normal	Decreased	Decreased
Higher center (cortical) lesions	Normal	Normal	Normal	Abnormal unless lesion is above Area 17

the flow of visual current. If all the testable levels, from retina to area 17, produce normal potentials, either the patient is malingering, or has a lesion higher than area 17, such as the patient reported by BODIS-WOLLNER [18].

Table 2 presents a brief outline of the results the clinician may obtain for several common eye diseases. Excluded from the table are indications of the latency findings for the various responses. Longer than normal ERG latencies are often associated with generalized degenerations of the retina, but there is at present not enough data on the focal or pattern ERG responses from abnormal eyes to characterize quantitatively the temporal aspects of the latter waveforms. Increased VEP latencies are, however, often associated with almost any condition affecting the flow of visual excitation which results in attenuation of VEP amplitude. This includes conditions as disparate as macular pathology and optic nerve neuropathy.

Summary

The basic rule, when using the flow charts for detecting and localizing visual abnormalities is to be mindful of the level of recorded activity. Lesions affecting cellular behavior at or close to the source (i.e., at the photoreceptors) necessarily impair the flow of visual information downstream from that level. A disturbance in the macular region may produce activity loss indistinguishable from a disease affecting the optic nerve. However, separate recordings of the focal ERG, pattern ERG, and VEP provide information from three major levels of current flow which greatly assists identification of the lesion site as well as providing a certain amount of quantitative data about the extent of damage and a clue to the nature of the disease itself.

References

1   R. Granit: J. Physiol. 77, 207 (1933)
2   G. Karpe: Acta Ophthalmol. Supp. 24, 1 (1945)
3   L.A. Riggs: Proc. Soc. Exp. Biol. N.Y. 48, 204 (1941)
4   H.M. Burian and L.A. Allen: EEG and Clin. Neurophysiol. 6, 509 (1954)
5   G. Goodman, H. Ripps and I.M. Siegel: Amer. J. Ophthal. 70, 214 (1963)
6   R.E. Carr and I.M. Siegel: Amer. J. Ophthal. 58, 95 (1964)
7   R.F. Miller and J.E. Dowling: J. Neurophysiol. 33, 323 (1970)
8   R.E. Carr and I.M. Siegel: Arch. Ophthal. 90, 327 (1973)

9   R.E. Carr, H. Ripps, I.M. Siegel, and R.A. Weale: Invest. Ophthal. 5, 497 (1066)
10 R.E. Carr and H. Ripps: Invest. Ophthal. 6, 426 (1967)
11 R.E. Carr, H. Ripps, and I.M. Siegel: Doc. Ophthal. Proc. Ser. IX ISCERG Symp., p. 193 (1974)
12 J.E. Dowling: Invest. Ophthal. 9, 655 (1970)
13 S. Sokol: Survey Ophthal. 21, 18 (1976)
14 L.A. Riggs, E.P. Johnson, and A.M. Schick: Science 144, 567 (1964)
15 A.M. Halliday, W.I. McDonald, and J. Mushin: Lancet 1, 982 (1970)
16 J.P. Kelly and D.C. VenEssen: J. Physiol. 238, 525 (1974)
17 M.J. Kupersmith, I.M. Siegel, and R.E. Carr: Arch. Neurol. 38, 362 (1981)
18 I. Bodis-Wollner et al.: Science 198, 629 (1977)
19 M.W. Brightman and T.S. Reese: J. Cell. Biol. 40, 668 (1969)
20 S. Sokol and B.H. Bloom: Doc. Ophthal. Proc. 13, 299 (1977)
21 G.B. Arden, Vaegen, and C.R. Hogg: Ann. N.Y. Acad. Sci. 388, 580 (1982)
22 G.B. Arden et al.: Invest.Ophthal. 18, 423 (1979)
23 W.W. Dawson, G.L. Trick and C.A. Litzkow: Invest. Ophthal. 18, 988 (1979)
24 G.B. Arden et al.: Trans. Ophthal. Soc. U.K. 100, 453 (1980)
25 L. Maffei and A. Fiorentini: Science 211, 953 (1981)
26 L. Maffei: N.Y. Acad. Sci. 388, 1 (1982)
27 W.H. Seiple et al.: Ophthalmol. 90, 759 (1983)
28 P. Wanger and H.E. Persson: Invest. Ophthal. 24, 749 (1983)
29 H. Asher: J. Physiol 112, 40P (1951)
30 T.S. Aiba, M. Alpern, and F. Maaseidvaag: J. Physiol. 189, 43 (1967)

Application of Laser Speckle Patterns to Visual Evoked Potentials

Hiroshi Uozato, Jun Fukuhara, Mototsugu Saishin, and Shuitsu Nakao

Department of Ophthalmology, Nara Medical University
Kashihara-shi, Nara-ken, 634, Japan

1. Introduction

Visual evoked potentials (VEPs) are gross electric signals generated in the occipital cortex in response to visual stimulation such as light or pattern. It is well known that the VEP is becoming a common clinical test for objective diagnosis of the function of the visual sense [1]. Recently versatile methods for the VEPs are commonly utilized, especially pattern VEPs which have been developed remarkably with the aid of microcomputer systems and commercial TV monitors [2]. However, these conventional pattern VEPs are strongly affected by refractive, accommodative, and fixation states of the subject's eyes. Although these methods are applicable to the objective determination of refraction [3], it is necessary to correct fully these factors for the assessment of visual function. In infants and patients with refractive or accommodative anomalies which are extremely difficult to correct, the conventional pattern VEPs are of little clinical value.

In this study, we propose a new method for the stimulation of VEPs using laser speckle patterns [4-6] which overcomes these drawbacks. The speckle patterns are focused on the subject's retina regardless of the refractive conditions of the eye [6-8]. Therefore, it is possible to obtain VEPs directly even in uncorrected ametropic eyes. This method is also applicable to the assessment of visual functions of cataractous eyes, which can hardly be performed by conventional methods [4-6,9].

In this paper, the principle of the speckle pattern stimulation for VEPs is briefly analyzed and some results of clinical applications are presented.

2. Principle and Method

When an optically rough surface is illuminated with highly coherent light, such as a laser, the scattered radiation possesses a marked granularity, or speckled appearance, and the images formed using this radiation also contain obvious speckle patterns. These speckle patterns are statistical in nature and their properties are intimately connected with both the statistical characteristics of the diffusing object and the coherence condition of the illuminating light [10-12]. The optical system of a laser speckle pattern stimulator employing a double-diffraction system and Maxwellian view optics [13] is shown schematically in Fig. 1. When a diffusing plate (D) having a circular aperture (diameter: w) is illuminated with a coherent laser light (wavelength: λ), the average speckle diameter at the subject's retina is given by

$$\sigma = \lambda \cdot f_1 \cdot fe / w \cdot f_2 \ , \tag{1}$$

where f_1, f_2 are the focal lengths of the double-diffraction lenses, and f_e

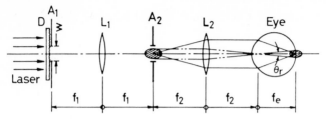

AVERAGE ANGULAR SPECKLE SIZE : $\theta_r = \dfrac{\lambda \cdot f_1}{w \cdot f_2}$ (rad.)

Fig. 1 Schematic diagram of the laser speckle pattern stimulator.
D: Diffusing plate. A_1 and A_2: Apertures. L_1 and L_2: Lenses.

is the distance from the second nodal point of the eye to the retina. Then
the average angular size of the speckle is expressed as

$$\theta_\gamma = \sigma/f_e = \lambda/w \cdot M \; , \qquad\qquad (2)$$

where M is the lateral magnification (f_2/f_1) of the double-diffraction system. Rewriting eq. (2) in order to express by spatial frequency (c/deg or
cpd), we get

$$\nu = 0.873 \times 10^{-2} \cdot w \cdot M / \lambda \; . \qquad\qquad (3)$$

From eqs. (2) and (3), it is found that the average speckle size or its spatial frequency at the retinal plane is independent of the optical system of
the eye. Since the variation of position of nodal and principal points due
to ocular accommodation is very small, we can neglect these effects; then the
average size at the retinal plane is constant regardless of the refractive
and accommodative conditions. Figure 2 shows the relation between average
spatial frequency of speckle pattern and diameter of illuminating area (w)
when the wavelength is 632.8 nm. In general, the cut-off frequency of the
visual system is the order of 50 cpd [14]; this range of spatial frequency
is easily covered by the variation of the illuminating area from 50 μm to
5 mm.

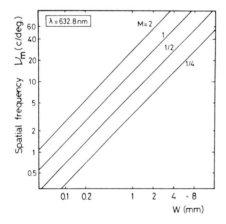

Fig. 2 Relation between average spatial frequency of the speckle pattern
and the diameter of the illuminating
area (w). M is the lateral magnification of the optical system

Another remarkable feature is the high contrast of speckle patterns. Theoretically, the average contrast of a speckle pattern is usually defined by a normalized standard deviation of speckle intensity variation at the observation plane as expressed in the next equation [10-12],

$$V = \frac{<\Delta I^2>}{<I>} = \frac{<I^2> - <I>^2}{<I>} \quad , \qquad (4)$$

where $< >$ stands for the ensemble average. When an ordinary diffuser is illuminated by a laser light, the next condition is satisfied,

$$<\Delta I^2> = <I^2> - <I>^2 = <I>^2 \quad . \qquad (5)$$

Therefore, the resultant contrast of a speckle pattern is almost unity. In our system, we can observe high contrast speckle patterns in spite of the refractive and accommodative states, or opaque ocular media.

The size of the stimulation field at the retina is controlled by the aperture size (2a) located at A_2 plane as shown in Fig. 1 and its diameter given by $\theta = 1/F_2$, where F_2 is the F number ($f_2/2a$) of the optical system.

As mentioned above, the average spatial frequency of speckle is controlled by the illuminating area of diffuser (w). However, the diffusing plane and the pupil plane are almost conjugate. When w becomes large enough, the pupil size itself determines the effective speckle size. Therefore, it is necessary to make the image of illuminating area (w) in the pupillary plane smaller than the actual pupil size. This condition is expressed as $w \cdot M \leq \emptyset$, where $\emptyset$ is the pupil diameter. The upper limits of the diameters of the illuminated areas for various pupil size is shown in Fig. 3.

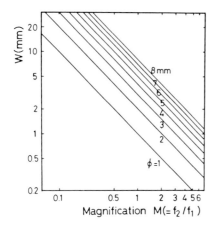

Fig. 3 Upper limit of the diameter of illuminating area (w) for various pupil sizes ($\emptyset$)

3. Experimental Results and Discussion

Modulation without luminance change can be performed by two methods, pattern shift and appearance-disappearance. Both methods are schematically shown in Fig. 4. He-Ne laser and tungsten light were used as coherent and incoherent light sources for contrast VEPs. A schematic diagram for recording and processing the potentials is shown in Fig. 5. VEPs are recorded differentially by two Ag-AgCl electrodes, one placed on the midline 3 cm above the inion and

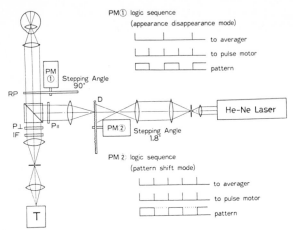

Fig. 4 Schematic diagram of a speckle pattern stimulator for appearance-disappearance and pattern shift modes. T: Tungsten light source. IF: Interference filter. RP: Rotating polarizer. $P_{||}$, $P_{\perp}$: Two orthogonal polarizers. D: Diffusing plate. PM: Pulse motor.

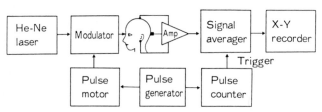

Fig. 5 Block diagram of the system used for recording and processing of VEPs.

the other on the right earlobe; the left earlobe serving as ground. The VEP signals are recorded as the average of 100 sweeps of 500 msec epochs, and the waveforms recorded with an X-Y recorder. All experiments were performed using binocular vision through Maxwellian view optics.

Influences of refractive state and accommodative fluctuation on the VEPs are shown in Figs. 6 and 7, respectively. Figure 8 shows the relation between VEP amplitude and visual acuity using logarithmic scales. VEP waveforms were also obtained from a patient with a corneal opacity and a patient with high myopia and cataract and are shown in Figs. 9 and 10, respectively.

One of the most excellent applications of the speckle pattern VEPs is in the case of opaque ocular media, such as a cataract [4-6], shown in Fig. 11. As already seen in Figs. 6-11, the conventional pattern VEPs (e.g., checkerboard patterns) are strongly influenced by the optical abnormalities of the eye such as refractive or accommodative problems, keratoconus, corneal opacity or cataract, etc. However, the speckle pattern method is recordable even in the presence of those latter abnormalities [8]. It is difficult, however, to obtain reliable VEPs in patients with dense and diffuse opacity such as a mature cataract even when interference fringe pattern VEPs [15,16] are used.

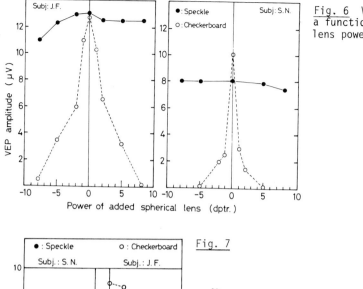

Fig. 6 VEP amplitudes as a function of spherical lens power

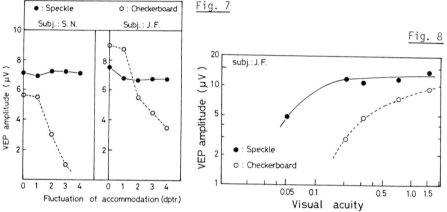

Fig. 7

Fig. 8

Fig. 7 VEP amplitudes as a function of accommodative fluctuation

Fig. 8 Relation between VEP amplitude and visual acuity in logarithmic scale

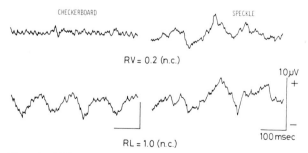

Fig. 9 VEP waveforms obtained from a patient with corneal opacity

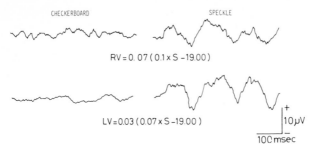

CHECKERBOARD   SPECKLE

RV = 0.07 (0.1 x S -19.00)

LV = 0.03 (0.07 x S -19.00)

+
10μV

100 msec

Fig. 10 VEP waveforms obtained from a patient with high myopia and cataract

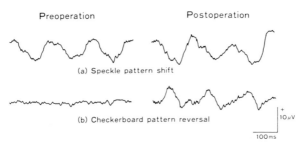

Preoperation   Postoperation

(a) Speckle pattern shift

(b) Checkerboard pattern reversal

+
10μV

100 ms

Fig. 11 VEP waveforms obtained from a cataractous patient. Speckle pattern shift (a) and checkerboard pattern reversal (b) VEPs are recorded from the same patient before and after operation, respectively.

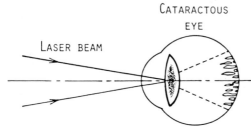

CATARACTOUS
EYE

LASER BEAM

Fig. 12 Application of the laser speckle pattern visual stimulation to cataractous eyes by using direct illumination of laser beam.

In this latter case, the speckle pattern VEPs can be detected, because the speckle pattern created by the cataractous lens itself is easily obtained by direct illumination of laser beam, as shown in Fig. 12 [4]. In general, the lens opacity is evenly distributed, and it is suspected that the average speckle size and its contrast is decreased by multiple scattering. Then the stimulation experiments shown in Fig. 13 are performed. Figure 14 shows the speckle patterns obtained with various grades of ground glass (a,b,c) and cataractous lens of a human eye in vitro (d). The results indicate that the finer the grain ground glass, the smaller is the average size of speckle but the contrast maintains a high value. Therefore, this technique may be useful even for a mature cataractous eye whose function is extremely difficult to ascertain with conventional pattern VEPs. It is of course important that laser safety regulation and the speckle size are given careful consideration.

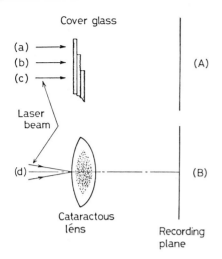

Cover glass

(a) ⟶
(b) ⟶
(c) ⟶

(A)

Laser
beam

(d)

(B)

Cataractous
léns

Recording
plane

Fig. 13 Schemes of the arrangement
used for recording diffraction-field
speckle patterns created with ground
glasses (A) and a cataractous lens
of a human eye (B)

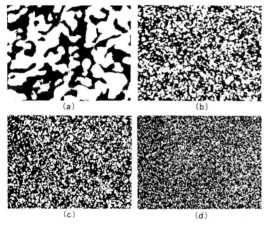

(a)

(b)

(c)

(d)

Fig. 14 Speckle patterns obtained with ground glasses
(a,b,c) and a cataractous lens
of a human eye in vitro (d)

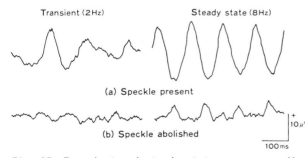

Transient (2Hz)   Steady state (8Hz)

(a) Speckle present

(b) Speckle abolished

$+$
$10\mu V$

$100\,ms$

Fig. 15 Transient and steady-state appearance-disappearance responses
obtained from a patient with a mature cataract using direct illumination
of laser beam.

171

Figure 15 shows the transient and steady-state appearance-disappearance responses obtained from a mature cataractous patient by using the direct illumination method of laser beam.

4. Conclusion

A new method of stimulation for obtaining visual evoked potentials (VEPs) uses laser speckle patterns. The optical properties of this stimulator are discussed and some experiments were performed to demonstrate the feasibility of speckle pattern VEPs. It is concluded that the speckle pattern VEP is very useful in the assessment of visual functions regardless of the refractive and accommodative anomalies, and even in the presence of opaque ocular media such as a cataract, corneal opacity, and so on. Therefore, the method can also be applied to infants with cataractous eyes whose visual function is all but impossible to obtain with conventional VEP methods.

Acknowledgments

The authors wish to thank Professor L.H. van der Tweel of the University of Amsterdam for his encouragement and helpful discussions. They are also grateful to A. Koike, M.D. and S. Nojima, M.D., for their technical assistance. This work was partially supported by a Grant-in-Aid for Scientific Research from the Ministry of Education (No. 577671 and No. 577673).

References

1 J.E. Desmedt (ed.): *Visual Evoked Potentials in Man: New Developments* (Clarendon Press, Oxford, 1977)
2 G.B. Arden, I. Bodis-Wollner, et al.: In *Visual Evoked Potentials in Man: New Developments*, ed. by J.E. Desmedt (Clarendon Press, Oxford, 1977), pp. 3-15
3 M. Millodot and L.A. Liggs: Arch. Ophthalmol. 84, 272 (1970)
4 H. Uozato, J. Fukuhara, and S. Nakao: J. Nara Med. Assoc. 31, 354 (1980)
5 H. Uozato, J. Fukuhara, and S. Nakao: J. Ophthalmol. Opt. Soc. Jpn. 2, 127 (1981)
6 H. Uozato, J. Fukuhara, M. Saishin, and S. Nakao: In *Optics in Biomedical Sciences*, ed. by G. von Bally and P. Greguss, Springer Series in Optical Sciences, Vol. 31 (Springer, Berlin, Heidelberg, New York, 1982), pp. 237-242
7 W.W. Dawson and M.C. Barris: Invest. Ophthalmol. Visual Sci. 17, 1209 (1978)
8 J. Fukuhara, H. Uozato, A. Koike, M. Saishin, and S. Nakao: Jpn. J. Clin. Ophthalmol. 35, 1009 (1981)
9 J. Fukuhara, H. Uozato, A. Koike, S. Nojima, M. Saishin, and S. Nakao: Acta Soc. Ophthalmol. Jpn. 85, 1559 (1981)
10 J.W. Goodman: In *Laser Speckle and Related Phenomena*, ed. by J. C. Dainty, Topics in Applied Physics, Vol. 9 (Springer, Berlin, Heidelberg, New York, 1975), pp. 9-75
11 I. Yamaguchi: Jpn. J. Optics (Kogaku) 3, 128 (1974)
12 T. Asakura: In *Speckle Metrology*, ed. by R.K. Erf (Academic Press, New York, San Francisco, London, 1978), pp. 11-49
13 G. Westhaimer: Vision Res. 6, 669 (1966)
14 T. Kawara and H. Ohzu: Oyo-Buturi 41, 128 (1977)
15 Y. Sugimachi: Acta Soc. Ophthalmol. Jpn. 82, 398 (1978)
16 G.B. Arden and U.B. Sheory: In *Visual Evoked Potentials in Man: New Developments*, ed. by J.E. Desmedt (Clarendon Press, Oxford, 1977), pp. 381-394

An Automatic Intermittent Light Stimulator to Record Flicker Persceptive Thresholds in Patients with Retinal Disease

J.J. Meyer, P. Rey, and A. Bousquet

Institute of Social and Preventive Medicine, University of Geneva
I-Geneva, Italy

S. Korol

University Eye Clinic, I-Geneva, Italy

1. Introduction

Modern life implies more and more the extensive use of vision. Although artificial lighting is extending at a faster rate, although visual displays are more and more invading the workplace, eye doctors do not know much about this particular variable which affects the normal as well as the pathological eye.

Are aging processes accelerated by visual overstrain and overstimulation which characterize the work area? Are ocular diseases aggravated by overstrain and overstimulation? Although ergophthalmologists and public health specialists are highly concerned about those problems, one may question whether the appropriate methods and tools exist to solve them.

On the other hand, numerous drugs have been demonstrated to be responsible for visual damage; a list of almost 100 of these substances has already been completed. These substances may well produce irreversible lesions either to the retina or to any other nervous structure along the visual pathways.

There is therefore an urgent need for techniques which would allow the early diagnosis of visual lesions by the physician, or even better with which the patient himself would be capable of detecting an early visual impairment either due to chemical therapy, or to his exposure to physical hazards.

This paper is aimed to describe such a technique.

2. Method

The investigating method makes use of flicker-fusion thresholds as plotted by de Lange (Fig. 1). When a subject is exposed to a sinusoidally modulated light, he can experience fusion for any given frequency under the critical fusion frequency, provided that the modulation amplitude of the light be settled at an appropriate value. Modulations are then plotted versus frequency, on a log scale. The de Lange curve displays a typical aspect with an abrupt attenuation of the high frequency region. We noticed that retinal diseases as well as pathologies of the optic nerve produced deviations in flicker-fusion thresholds from normal values. Let us mention that BREUKING [1] had described clinical cases with changes in flicker sensitivity. More recently, we demonstrated how useful it was to compare flicker ERG to flashes to threshold perceptive flicker for the early detection and follow-up of diverse macular pathologies [2,3,4,5].

These results were promising. But we were conscious that before it could be accepted as a good candidate for the everyday clinical and field applica-

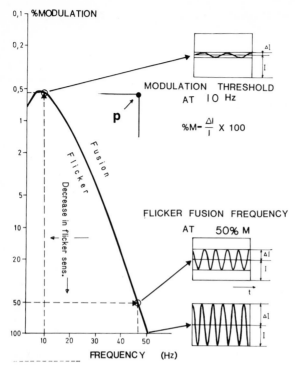

Fig.1. A typical de Lange curve divides the diagramm in two areas: the one under the curve where light flickers and the one over where fusion occurs. A decrease in flicker sensitivity corresponds to a depression of the curve following the 2 arrows. On the right of the figure, representation of the modulation amplitude at three frequencies: P is the projection of the two thresholds which were selected to express the main changes of the curve (Simplified Attenuation Characteristics. S.A.C)

tion, we still had to improve the test. First, the threshold determination was much too dependent on the operator's skill. Second, the test surface being too small, we were not capable of recording retinal responses outside the perifoveolar region.

2.1 Apparatus and Procedure

Different techniques have been used to generate sinusoidally modulated light to stimulate the eye and to record the thresholds which are necessary to establish the attenuation characteristics of the visual system. Each of them has its advantages and disadvantages. For example, our original finding was to make use of a well-defined contrast between a one-degree test area and a 30-degree background surface surrounding it. Most of our experiments were at first performed with a TV tube providing a spot of one degree which was seen in front of a background surface indirectly illuminated by incandescent lamps [2,3,4].

Our new method differs from the older one in several respects:

1) To suppress any color contrast effect, we adopted for background and test light sources the same fluorescent lamp (TL 55 de luxe) (Fig. 2) [6].

2) Using indirect lighting we could illuminate any given white surface on a dark background. After some preliminary experiments, we chose the three following situations:
   - First condition (A): one degree flickering area at a mean luminance of 200 cd/m^2, bright surround at a luminance of 300 cd/m^2;
   - Second condition (B): six degrees flickering area at 5 cd/m^2, dark surround at 0.5 cd/m^2;
   - Third condition (C): 30 degrees flickering area at 25 cd/m^2, remaining visual field in the dark.

3) We elaborated a semi-automatic recording procedure with a direct write-out of the curve by means of an X-Y plotter. The frequency of stimulation is automatically increased by 5 Hz after each modulation threshold is determined. The range of frequencies goes from 5 Hz to 75 Hz.

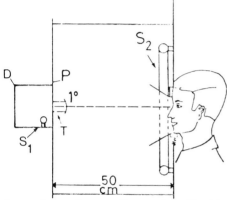

Fig.2. Frontal view of the display. D: Background panel reflecting the light coming from the lamp S1 and stimulating the eye through an aperture of one degree. P: panel reflecting the light provided by two fluorescent lamps (S2) in the whole visual field. The amount of reflected light is changed, by changing the reflective power of P and D

Here, in brief, is a description of the measurement procedure. During the preadaptation period (3 to 5 minutes), the subject is instructed to fix a given region of the panel and to press a button when he sees flicker, until flicker disappears, and to repeat the same procedure when flicker appears again. Then, the subject is required to practice for a few more minutes during which the operator can check whether or not the subject is able to perform the test properly. Pushing a knob, the operator initiates the automatic procedure. When the whole curve is recorded, a slowly flickering signal indicates that this is the end. After a few minutes recovery, the subject is required to repeat the test.

2.2 A Simplified Method to Plot Flicker Thresholds

We showed that the main changes in the form of the de Lange curve could be expressed by means of two thresholds only: one representing the left side of the curve and one representing the right side (Figs. 1 and 3). The first one is the threshold modulation amplitude at 10 Hz; the second one is the fusion frequency for a modulation amplitude of 50%.

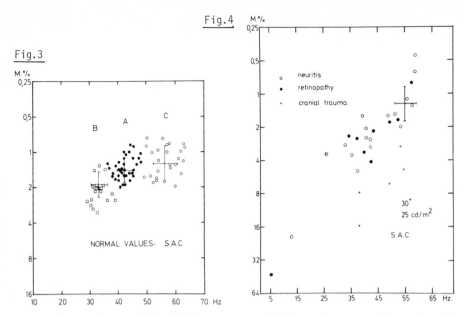

Fig.3

Fig.4

Fig.3. P coordinates (S.A.C.) for a group of 26 normal persons investigated in 3 different conditions A,B, C as described in text.

Fig.4. P coordinates for pathological cases (condition C). Mean and standard deviation for normal eyes. One sees that P values move more and more from the normal area when the lesion becomes more severe; no difference can be expected with the location of the lesion, outside perhaps between peripheral and central locations.

3. Results

3.1 The Effect of Changing the Stimulus Intensity or Test Area on Normal Subjects (Fig. 3)

An increase in test intensity or area will produce, mostly, a shift of the right side of the de Lange curve which means an increased sensitivity for higher frequencies. With age, there is a decrease in flicker sensitivity (Table 1) which is particularly noticeable in central fixation and photopic adaptation. When the test spot is larger, the effect of aging can be masked or even reversed.

In young people, we demonstrated a depression of flicker sensitivity in the case of vitamin A or retinal binding protein deficiency [7].

From these results, it can be concluded that the knowledge of testing conditions (stimulus intensity, area, adaptation level and so on) is necessary in order to interpret and quantify variations of de Lange curves.

3.2 Pathological Cases

Since our method is non-invasive and very simple, it can be applied very easily to patients provided that the appropriate speed and timing of the automatic procedure is used.

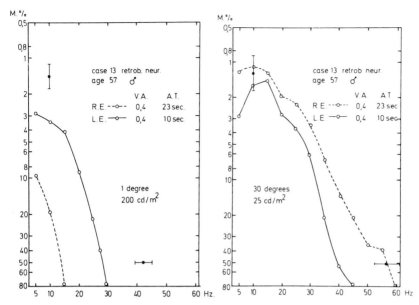

Fig.5a,b. Change in the shape of the de Lange curve with the change of stimulation (intensity and area) for two eyes.

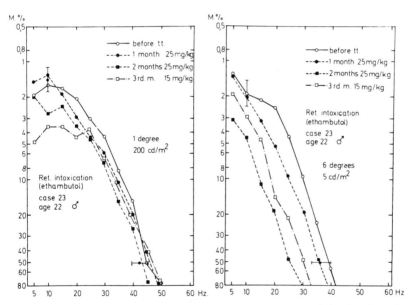

Fig.6a,b. Change in the shape of the de Lange curve with the duration of intoxication. One sees that those modifications are emphasized by condition B in comparison with condition A.

Table 1 Simplified attenuation characteristics (S.A.C.) reference values for high frequency region. Mean and standard deviation (N = 33 young and 44 aging normal subjects) of frequency thresholds at 50% modulation amplitude, as measured in three conditions A, B, and C. A significant decrease with age was found. Mean values expressed in cycles per second (Hz).

Test Condition	A 1°/200 cd/m^2	B 6°/ 5 cd/m^2	C 30°/25 cd/m^2
44 young subjects (18-25 years)	43.6 ± s 2.4	33.6 ± s 3.6	56 ± s 4.4
35 older subjects (26-45 years)	40.8 ± s 2.8	33.2 ± s 2.8	54.8 ± s 4.8
age effect	*** - 2.8	- 0.4	- 1.2

*** P < 0.0005

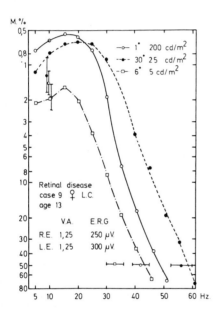

Fig.7. Abnormal enhancement of flicker sensitivity, as measured in 3 conditions (A, B and C) in an early detected case of retinitis pigmentosa

Three pathological cases (retinopathy, retrobulbar neuritis, and cranial trauma) are described in Figs. 4, 5, 6, and 7. They represent three levels of the visual pathways. The importance of the shift of pathological curves from normal depends very strongly on stimulating conditions, as can be seen in comparing Fig. 5A and 5B. Such a difference is particularly noticeable in one case of ethambutol intoxication (Fig. 6A and 6B). With a 6° test area and a low luminance level, the depressing effect of ethambutol with time is very much enhanced compared to a 1° area and photopic adaptation. During the same period, control subjects were followed up but no change in flicker sensitivity was detected. Let us stress here that toxic substances may produce at first an enhancement of flicker sensitivity [4,5]. A similar increase could be due to a starting pathological process, as shown by Fig. 7.

4. Conclusions

1. Measuring flicker fusion thresholds is an appropriate method for the early detection of retinal disturbances since it is not complicated by either refraction defects or morphoscopic vision.

2. When comparing a threshold value to some reference value (either from the same individual or from a normal group), a shift in flicker fusion sensitivity may be considered as revealing some retinal or retrobulbar involvement. As such, it can be applied to early detection as well as to follow up.

3. In association with flicker ERG, our method will be able to point out the target-organ as in the case of ethambutol intoxication which may affect either the nervus opticus or one or the other retinal layer.

4. Since our method is non-invasive and provides a quantitative evaluation of the severity of the retinal involvement, it should be used to reduce the number of times an ERG is recorded, thus avoiding discomfort and perhaps additional risk to the patient, particularly in children.

5. Finally, in case of unexplained sensitivity to glare, flicker fusion thresholds could be shown to be disturbed while results of routine glare testing remain unchanged.

In summary, our technique is particularly useful for the early detection, diagnosis and follow-up studies of retinal disease. We believe it can serve as a useful tool for ophthalmology and preventive medicine.

Acknowledgments

The valuable assistance by L. Zoganas and J. Pittard is gratefully acknowledged. We thank the firm of Hoffmann-La Roche for the research grant.

References

1 E.W. Breukink: Ophthalmologica 146, 143 (1963)
2 J. Babel, P. Rey, N. Stannos, J.-J. Meyer, and P. Guggenheim: Doc. Ophth. 26, 248 (1969)
3 J.-J. Meyer, P. Rey, B. Thorens, and A. Beaumanoir: Archives suisses de Neurologie, Neurochirurgie et de Psychiatrie 108, 213 (1971)
4 J.-J. Meyer, S. Korol, and J. Babel: Archs. ophthal., Paris 35, 253 (1975)
5 J.-J. Meyer, S. Korol, R. Gramoni, and R. Tupling: Mod. Probl. Ophthal., Vol. 19 (Bâle, 1978), pp. 33-49
6 J. Richez and J.-J. Meyer: Med. and Biol. Eng. and Comput. 15, 298 (1977)
7 J.-J. Meyer, A. Bousquet, S. Korol, and G. Brubacher: Revue de med. soc. et prév. 26, 310 (1981)

Superacuity in the Spared Eyes of Monocular Deprivation Amblyopes: Visual Evoked Response Measurements

George B. Jastrzebski and Elwin Marg

School of Optometry, University of California, Berkeley, CA 94720, USA

Creig S. Hoyt

Department of Ophthalmology, University of California, San Francisco, CA, USA

Introduction

Studies of the effects of depriving visual environments on the young visual systems of-cats and monkeys have led to the definition of a "critical period" during which such deprivation results in losses of visual system function and morphology [1]. The reversal of monocular deprivation in animals by means of reverse deprivation forms the basis for the clinical treatment of deprivation amblyopia. In man, patching most commonly serves as the analogous procedure for lid suture in the animal studies. In cats and monkeys, through what is believed to be a process of binocular competition for neural connectivity the originally deprived eye recovers as a result of reverse deprivation, while the newly deprived eye suffers the usual effects of deprivation. Such changes occur only during the "sensitive" period [1,2]. Recently, similar changes in the visual acuity of young human infants undergoing patching have been observed by using a forced-choice preferential looking (FPL) method [3] and visual evoked responses (VER) [4,5,6]. As is predicted by the animal models the amblyopic eye gains acuity at the expense of the normal. Binocular competition could explain this symmetrical push-pull type of effect on either ocular dominance as measured in animals or visual acuity as measured in human infants. Recently, measurement of visual acuity by means of preferential looking or visual evoked response has made possible the "titration" of the patching to optimize the therapeutic result by avoiding unnecessary monocular deprivation which is clearly detrimental to binocularity and minimizing the risks of reverse or occlusion amblyopia. Such work has also shed some light on the processes underlying normal and abnormal visual development in humans.

We have proposed and demonstrated [6,7] a unique model for deprivation amblyopia which specifically addresses clinical aspects of stimulus deprivation and simultaneously reconciles animal and human data. The model describes amblyopiogenesis from both natural deprivation and occlusion. Our model (called the SPC model) is based on assigning three properties to visual acuity in the young visual system: sensitivity, plasticity, and elasticity. Sensitivity allows the acuity to decrease (ipsilateral effect) or increase (contralateral effect) during deprivation. When deprivation is removed elasticity allows for spontaneous reversal of these changes while plasticity denotes deprivation effects that do not reverse spontaneously upon removal of the deprivation. If the effects of monocular deprivation are equal but opposite in the two eyes, we say that the system is symmetrical and binocular competition is preserved as the underlying mechanism for the changes seen, be they acuity or ocular dominance. Simply, the making of connections for one eye is the breaking of connections for the other. At this point our model predicts that occlusion will only improve vision in an amblyopic eye at the expense of acuity in the normal. This takes into account "reverse" or "occlusion" amblyopia.

In satisfying requirements for successful occlusion the model must provide for asymmetry in sensitivity, plasticity, and elasticity between the normal and amblyopic eyes. While this requires a discounting of the binocular competition mechanism (new connections for the amblyopic eye without breaking connections for the normal), successful occlusion therapy of amblyopia is accounted for. We have begun to determine S, P and E in a quantitative manner in both eyes of young amblyopes incidentally to the normal course of occlusion treatment. Data collected from this preliminary study bear out both symmetric and asymmetric schemes of our model. The degree of symmetry of the response to patching appears to be related to the severity of the deprivation (hours per day in this case). Extremes of patching or discontinuation of patching produce symmetrical changes in acuities, while intermediate levels of patching produce largely asymmetrical changes in acuity. It appears then, that intermediate levels of patching may be more appropriate for successful occlusion. This agrees with animal data that shows reverse experience is more effective when broken up into a series of short exposure sessions [8]. A learning mechanism is suggested for this effect. The SPE model postulates that during the development of acuity the formation of such putative "new" connections is disturbed by monocular deprivation and the development of acuity is shifted to the spared eye. A mechanism of binocular competition is sufficient to explain this effect. Our model predicts that monocular deprivation causes not only a loss of acuity in the deprived eye, but a gain of acuity in the spared eye. The spared eye can be expected to have an acuity greater than the acuity normally expected for the developmental age. It also predicts that such paradoxical superacuity would no longer be apparent in an individual once the visual acuity reaches the adult maximum. This may be due to saturation of the acuity by a limiting factor of resolution, for example the ocular optics or the density of the retinal mosaic [9].

MARG et al. [10] describe the baseline for normal development of acuity in infants as measured by transient pattern onset/offset visual evoked responses. Adult acuity of 20/20 equivalence is demonstrated in infants at about 6 months of age. The baseline for preferential looking acuities is somewhat lower with adult level acuity found at age 12 months [11]. This report describes a preliminary series of VER acuity measurements in a group of children experiencing or having recently experienced visual deprivation during the period of acuity development. The anatomical evidence in cats and monkeys would suggest a conservative mechanism underlying monocular deprivation. Cortical representation increases for the spared eye at the expense of the deprived in keeping with the binocular competition model. The anatomical evidence favors a mechanism of recruitment as opposed to atrophy to explain monocular deprivation and one might expect to find enhancement of acuity if one could discount the limiting factors of diffraction optics and retinal mosaic. FREEMAN and THIBOS [12,13] could not find such enhancement of function between orientation channels in meridional amblyopia the consequence of a more selective deprivation, astigmatism. FREEMAN and BRADLEY [14] find that the spared eyes of monocular deprivation amblyopes demonstrate enhancement of vernier acuity, a higher order spatial vision processing. This effect appears, as predicted, only in functionally monocular individuals, but the authors admit that this may well be an artifact of practice. The current study reports on the visual acuity as determined by transient pattern VER measurements of the "normal" eye in a variety of subjects demonstrating clearly several mechanistic features of superacuity. MOHINDRA et al. [15] briefly report a superacuity among monocularly deprived individuals using a forced choice preferential looking technique. Although several salient similarities exist between their and the present reports, a number of discrepancies are examined below.

Methods

The subjects in this study were selected among patients seen at the Children's Eye Clinic at the University of California, San Francisco School of Medicine and are listed and described in Table 1. Subjects 1-6 were monocularly deprived and have no history of patching. Subject 7 was monocularly deprived and has a history of intermittent 4 hours per day patching. Subjects 8 and 9 were monocularly deprived and undergoing full-time patching. Subject 12 was a strabismic amblyope undergoing patching. Acuities were determined as follows [4]: subjects were held in the parent's lap approximately 1.1 meters from a translucent rear projection screen that subtended about 20 degrees. An appearance/disappearance checkerboard stimulus was generated from a pair of optically coincident slide projectors. Electromechanical shutters limited presentation of the patterned target to 40-50 msec. replacing the pattern with a uniform grey field of equal mean luminance for the remainder of the interstimulus period. Transient evoked responses were recorded from scalp electrodes placed at 1 cm above the inion (active) and right and left earlobes or mastoid processes (reference and ground respectively). Amplification of 100,000 times was provided by two Grass P-14 amplifiers in cascade each with a bandwidth of 1 to 30 Hz. The average waveform to at least 64 presentations at 1.0 to 1.5 presentations per second was obtained by a signal averager (Nicolet). Several checkerboard sizes were presented and acuity was determined from a regression of amplitude versus log spatial frequency. Subjects wore soft contact lens optical correction for the test if necessary.

Table 1 Subjects for Superacuity study. Snellen equivalent visual acuity is given for each subject for the better eye. Round symbols represent subject's deprivation conditions as: Open circle - Strabismus; Half-filled circle - monocular deprivation; Single bar to left of circle - part-time patching; Double bar - full-time patching.

SUBJECT		AGE	ACUITY	CONDITIONS
1	◐	3 w	20/80	mon. cataract
2	◐	3 w	20/120	mon. absent pupil
3	◐	2 m	20/60	mon. cataract
4	◐	2-1/2m	20/70	mon. cataract
5	◐	2-1/2m	20/120	mon. ptosis
6	◑	3 m	20/60	mon. cataract
7	l◐	5 m	20/40	m. cat. p-t patch
8	ll◐	8 m	20/60	m. cat. f-t patch
9	ll◐	9 m	20/25	m. cat. f-t patch
10	●	2-1/2m	20/200	bilat. cataracts
11	●	4 m	20/40	bilat. cataracts
12	○	6 m	20/80	strabismus LXT

Results

The acuities of the normal eye in each of the subjects are presented in Fig. 1, superimposed onto baseline curves for normal acuity development based on VER and Preferential Looking techniques (curves A and B respectively). These

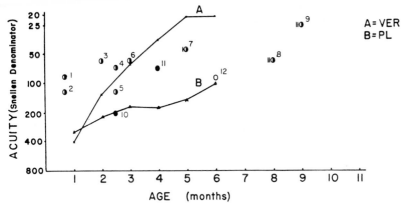

Fig. 1 Visual acuity of the better eye in deprivation amblyopes. The ordinate is visual acuity (Snellen equivalent denominator: 20/Y) on a logarithmic scale. The abscissa is age at time of testing in months. Curve A represents acuities for normal infants as determined by VER - visual evoked response [10]. Curve B indicates visual acuities for normal infants as determined by PL - preferential looking [11]. Subjects are labeled by number.
Symbols as in Table 1.

baselines are composites of normal acuities in figures from MARG et al. [10],
and DOBSON and TELLER [11]. Monocularly deprived subjects that have not been
patched (1-4 and 6) show a superacuity relative to the normal acuities given
for the VER technique. This effect clearly diminishes with age as can be
seen from the small effect in subject 6 (3 months) versus subject 1 (3 weeks).
Subjects 7, 8, and 9 show a lower than normal acuity. This is probably the
result of these subjects' patching history. Binocularly deprived subjects
(10 and 11) show a less than normal acuity (the better eye) even before the
period for developing 20/20 acuity is complete (6 months). Apparently, binocular deprivation has some detrimental effect on acuity in both eyes. The
strabismic (subject 12) also shows a less than normal acuity presumably due
to patching. Subject 5 fails to show superacuity despite monocular deprivation and appropriate age (compare with subject 4). Superacuity is plotted
versus age in Fig. 2. A regression line is fit for subjects below the age
of 3.5 months with two exclusions. Subject 10 is excluded being a binocularly deprived individual. Subject 5 is excluded as an outlier where the form
of deprivation was different. Subject 2 was retained in the regression, but
also appears to be an outlier. The regression indicates a period lasting

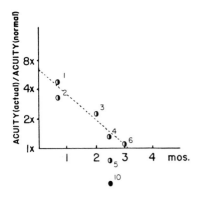

Fig. 2 Superacuity ratio in younger subjects 1-6 and 10 (less than 3.5 months)
showing acuity in excess of the normal VER
data. The ordinate is in octaves determined
as the ratio of actual acuity (reciprocal of
Snellen denominator) and expected acuity
(Snellen denominator from normal VER data -
curve A). Abscissa is age in months. The regression line shown includes all of the
points in Fig.2 except subjects 5 and 10.
The Superacuity ratio is 1 (no effect) at
about age 12.5 weeks or 3 months (r = 0.96)

until 12.5 weeks. The exclusion of these two subjects results in a correlation coefficient of 0.96.

Discussion

The appearance of superacuity during the period of acuity development provides some insights into the mechanisms for deprivation amblyopiogenesis. That superacuity was not demonstrated in binocularly deprived individuals is evidence for a binocular competition type mechanism for both amblyopia and superacuity. During binocular deprivation the stress upon the visual system subserving each eye is approximately equal. Binocular deprivation does not produce the competition that monocular deprivation does for neural connectivity in the visual pathway and this is reflected in the animal model [1]. Our binocularly (cataract) deprived individuals show a somewhat lower acuity in both eyes and such a result is also reported by JACOBSON et al. [3]. The push-pull model described in the symmetric actions of the SPE model predicts no shift of acuity from one eye to the other. However, if the deprivation is monocular, the stress is unbalanced and the nondeprived eye will gain acuity at the expense of the deprived. The appearance of superacuity is limited to the monocular deprived subjects and this is correctly predicted by the binocular competition mechanism. MOHINDRA et al. [15] state that individuals with monocular microophthalmia or anophthalmia do not show superacuity. The suggestion is that some level of visual input is still required for the interocular interaction that results in superacuity. Monocularly deprived subjects which have undergone substantial patching show the expected loss of acuity in the "normal" eye. It is possible that superacuity goes undetected because early patching, often instituted in cases of monocular deprivation, may readily wipe out the entire effect. The binocular competition mechanism is represented by symmetric acuity changes in the occlusion situation and is associated with extremes of full-time occlusion or discontinuation of occlusion. It is anticipated that the mechanism that explains deprivation amblyopiogenesis will maintain this underlying mechanism since naturally occurring deprivation is usually (at least in the case of compromised optic media) a full-time deprivation. Subject 5 (monocular deprivation from ptosis) fails to show superacuity perhaps because of the difference in the severity of the deprivation when compared to the congenital cataract cases.

The waning of superacuity with age is expected in the light of the concept of a critical period during which acuity is developed and appropriate neural connectivity is established. The belief that intervention of congenital monocular cataract before the age of 6 months [16] is consistent with preventing the competitive process of deprivation from persisting through the period of acuity development. Amblyopiogenesis during this developmental period is particularly resistant to treatment since the redistribution of "connections" will require the asymmetric processes described in the SPE model. This may be the mechanistic basis for the amblyopia of arrest of WORTH and CHAVASSE [17]. Amblyopiogenesis occurring after the development of acuity can be expected to be more readily reversed since "new connections" are not involved in the acuity changes of the deprivation. This may represent the amblyopia of extinction [17] and the easily reversible occlusion amblyopia [18]. It is curious that the superacuity wanes well before the end of acuity development (3 months versus 6 months). Data from infant monkeys would suggest that development of the optical properties is probably not a major limiting factor to the development of acuity [19]. Extension of this data to humans would remove visual optics as the limiting factor to superacuity.

Comparing the MOHINDRA [15] study to the present report is difficult since the two measures of acuity VER and FPL do not fully agree in absolute terms on the course of acuity development. The possible reasons for disagreement are complex and defy explanation here except to say that adult levels of acuity by VER are reached by about age 6 months at which age FPL estimates acuity at 20/100. FPL continues to measure a superacuity through nearly 40 weeks of age and a regression of superacuity would extend superacuity to well past the first year of life. Further, acuity levels reported as superacuity under the FPL criterion fall on or below normal acuity levels for the VER criterion (line A in Fig. 1). This disparity in criteria for superacuity between the two acuity measures does not necessarily compromise the validity of either observation of superacuity. HELD et al. [20] report that FPL may tend to underestimate acuity since targets with very fine detail are discriminated by, but not preferred by infants. It is possible that monocular deprivation unmasks this avoidance of detail and produces a superacuity. It is also possible that behavioral responsiveness is enhanced when the normal eye is uncovered and that FPL superacuity follows a time course contingent on the development of higher behavioral centers. The VER relies only on the responsiveness of the central visual pathway. This partially explains the higher estimates of acuity, and also suggests a more direct physiologic link for the acuity measure.

One might argue that the reason for our being able to demonstrate superacuity in the current study follows from testing at a period in life where the resolution limit of optics and retinal mosaic have been made remote. We feel this certainly contributes to the visibility of the effect. The time course of the effect, however, suggests a second limiting factor has been approached. If indeed the optical and retinal factors were still limiting superacuity one might expect superacuity to last until adult acuity, limited by these factors, is attained. The distinction which allows for the disproportionate formation of connections, if you will, on the nondeprived side, perhaps manifesting itself here as superacuity, may be the presence of "uncommitted" synapses, etc., during early development [21]. It may be for this reason that superacuity ends at approximately the same time that binocularity and stereopsis begin to develop [22,23]. The "extra" neuronal pool may be diminished due to the consumption by the genesis of binocularity and perhaps other higher-order processes such as vernier acuity and stereoacuity. The potential for correcting "amblyopia" by competition is also diminished and the deprivation becomes more permanent. Superacuity is also expended. This is supported by the restriction of enhanced vernier acuity to functionally monocular individuals. The present study in using such young subjects minimizes the complication of a practice effect.

Of course, the relationship of superacuity to any of these neuronal mechanisms is speculative and further refinement of these measurements of superacuity will be necessary to document adequately the phenomenon and its underlying conditions. The site(s) of neuronal mechanisms underlying deprivation are yet to be unequivocally determined. Our measurements in human infants do suggest, however, that much that is not only of clinical value, but of theoretical import can be gained from the study of this population in this way.

Conclusions

Demonstration of the SPE model's prediction for the occurrence of superacuity as a consequence to binocular competition during monocular deprivation in young infants corroborates several observations:

1) The occurrence of superacuity during the development of acuity. This is acuity in the spared eye above the normal level for the developmental age.

185

2) The occurrence of superacuity only in monocularly deprived individuals since binocular deprivation stresses the system from both sides resulting in symmetric visual development.

3) The observation that naturally occurring deprivation involving the optic media is generally full-time and the mechanism of binocular competition is consistent for the analogous occlusion situation: full-time occlusion or nonocclusion produces symmetric acuity effects.

4) The SPE model reflects the classical clinical impressions of amblyopia of arrest and extinction while remaining consistent with the more recent human and animal data.

5) The concept of active neural recruitment due to monocular deprivation is supported by anatomical changes suggestive of functional development.

6) Superacuity is helped in becoming manifest by the choice of very young subjects in whom optical and perhaps retinal resolution limits are more remote from neural limits. Also perhaps, competing higher processes such as binocularity are not yet developed, allowing superacuity to become apparent.

References

1   D.H. Hubel and T.N. Wiesel: J. Physiol. $\underline{206}$, 419 (1970)

2   C. Blakemore and R.C. Van Sluyters: J. Physiol. $\underline{237}$, 195 (1974)

3   S.G. Jacobson, I. Mohindra, and R. Held: Am. J. Ophthal. $\underline{93}$, 198 (1982)

4   J.V. Odom, C.S. Hoyt, and E. Marg: Arch. Ophthal. $\underline{99}$, 1412 (1981)

5   J.V. Odom, C.S. Hoyt, and E. Mart: Am. J. Optom. Physiol. Optics $\underline{59}$, 706 (1982)

6   G.B. Jastrzebski, C.S. Hoyt, and E. Marg: Stimulus deprivation in young children I: The SEP model. In press

7   G.B. Jastrzebski, C.S. Hoyt, and E. Mart: Stimulus deprivation in young children II: Measurements of S.E.P. In press

8   D.P. Crewther, S.G. Crewther, and D.E. Mitchell: Invest. Ophthal. $\underline{21}$, 357 (1981)

9   F.W. Campbell and D.G. Green: J. Physiol. $\underline{181}$: 576 (1965)

10 E. Marg, D.N. Freeman, P. Peltzman, and P. Goldstein: Invest. Ophthal. $\underline{15}$, 150 (1976)

11 V. Dobson and D.Y. Teller: Vis. Res. $\underline{18}$, 1469 (1978)

12 R.D. Freeman and L.N. Thibos: J. Physiol. $\underline{247}$, 687 (1975a)

13 R.D. Freeman and L.N. Thibos: J. Physiol. $\underline{247}$, 710 (1975b)

14 R.D. Freeman and A. Bradley: J. Neurophysiol. $\underline{43}$, 1645 (1980)

15 I. Mohindra, R. Held, and S.G. Jacobson: (Abstract to ARVO, April 1981, Sarasota, Fa.) Suppl. Invest. Ophthalmol. & Vis. Sci. $\underline{20}$, 119 (1981)

16 R. Beller, C.S. Hoyt, E. Marg, and J.V. Odom: Am. J. Ophthal. $\underline{91}$, 559 (1981)

17 C. Worth: In *Worth and Chavasse's Squint; The Binocular Reflexes and the Treatment of Strabismus*, 9th ed., T.K. Lyle and G.J.O. Bridgeman (London, Bailliere, Tindall & Cox, 1959)

18 G.K. von Noorden: Am. J. Ophthal. $\underline{69}$, 210 (1970)

19 R.A. Williams and R.G. Boothe: Invest. Ophthal. & Vis. Sci. $\underline{21}$, 728 (1981)

20 R. Held, J. Gwiazda, S. Brill, I. Mohindra, and J. Wolfe: Vision Res. $\underline{19}$, 1377 (1979)

21 P. Rakic: Nature $\underline{261}$, 467 (1976)

22 R. Fox, R.N. Aslin, S.L. Shea, and S.T. Dumais: Science $\underline{207}$, 323 (1979)

23 R. Held, E. Birch, and J. Gwiazda: Proc. Natl. Acad. Sci. $\underline{77}$, 5527 (1980)

Pupillary Escape and Visual Fatigue Phenomena in Optic Nerve Disease

Satoshi Ishikawa and Kazuhiko Ukai

Department of Ophthalmology, School of Medicine, Kitasato University
Sagamihara, Kanagawa 228, Japan

1. Introduction

An afferent pupillary defect has been known as an important sign to detect abnormality of the optic nerve such as optic neuropathy. Clinically, light swing test is the most common method to detect the afferent pupillary defect. Recently, pupil cycle time (PCT) has been measured by using a stop watch under the standard slit lamp microscope [1]. When a small beam is focused on the pupillary margin, it provokes regular and persistent oscillation of the pupil. The PCT is the frequency of this oscillation converted in msec. Prolongation of PCT has been suggested in patients with retrobulbar optic neuritis (RBN) when measured under the slit lamp microscope [2]. Since the technique is relatively simple, we attempted to measure the PCT with RBN by using the same procedure. The results, however, differed according to the examiner, from patient to patient, and it was quite difficult to obtain a consistent PCT in a given patient. We have, therefore, recorded edge-light pupil oscillation by using a small slit of light projected by modifying streak retinoscope, and the pupil was observed by a sensitive infrared pupillometer [3]. The PCT was initially calculated by hand; thereafter, a digital computer with analogue to digital converter was used for calculting the PCT by Fourier analysis in normal people and in patients with RBN. The results indicated that there were no significant increases in the PCT in patients with optic neuropathy by both methods. On the other hand, pupil escape phenomenon was frequently seen during edge light pupillary oscillation in these patients [4].

In the present study, pupillary abnormality, i.e. pupil escape, was seen in the patients with RBN when observed by edge—light pupillary oscillation and by intense light stimuli.

2. Methods

Two experimental methods were employed in this study. A detailed description has been made elsewhere [4], therefore a brief description will be made.

2.1 Edge-Light Pupillary Oscillation

In order to produce an edge-light pupil oscillation (ELPO), the following method was used. For the stimulus, a streak retinoscope (Neitz Model RI), whose mirror head had been removed, was used. The distance between the retinoscope and the eye was 15 cm. A dichroic mirror was placed between the eye and the retinoscope. In this manner, the pupil image was displayed on the surface of an infrared (IR) television (TV) camera. The signal from the camera was fed into an amplifier (Iriscorder Model C-301: Hamamatsu TV) and the output was connected to a graphic recorder.

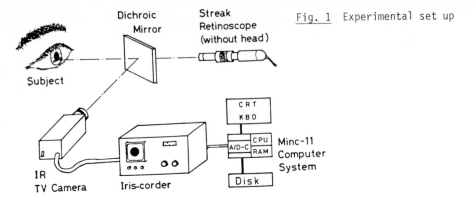

Dichroic Mirror Streak Retinoscope (without head)

Fig. 1 Experimental set up

Subject

CRT
KBD
A/D-C | CPU RAM   Minc-11 Computer System
IR TV Camera   Iris-corder   Disk

Apparatus

An analogue signal was recorded on the recorder. If the oscillation was regular and continuous, a period of at least three minutes was recorded in each case. About 100 regular oscillations were selected and the PCT expressed in msec was obtained. If the oscillation was irregular, at least five continuous oscillations, selected from among the most regular oscillations on the recorder, were chosen. In all cases, these oscillations were obtained during the course of ten sessions and using them, the average PCT was determined.

In the second experiment, we introduced a Minc-11 computer by connecting the output from the pupillometer. The power spectrum was obtained using the fast Fourier transformation subprogram to determine the dominant frequency. The data from at least ten sessions within 20 seconds were examined and then averaged.

Sixty healthy subjects were used as controls. The patients group (40 cases) was selected from among individuals diagnosed as cases of optic neuritis or retrobulbar optic neuritis at the neuro-ophthalmology clinic at Kitasato University.

2.2 Intense Light Stimulation

After 15 minutes dark adaptation, the eye was illuminated by intense light (10,000 cd/m^2) for 10 seconds followed by 1 second dark interval. This was designated as one session. Ten sessions were repeated and pupillary movement during above stimuli was computed. For the calculation, the Minc-11 computer was also employed. Twenty three cases out of 40 cases were selected and used for this study.

3. Results

3.1 Analogue Data Analysis

Normal subjects: Typical pupil oscillations obtained from four normal subjects are shown in Figure 2. Their age and sex are indicated at the left, and the calibration of the pupil area with the time scale at the right. There are slight variations among the subjects in the amplitude and frequency of the oscillation. With older subjects, the amplitude tended to decrease. But the baseline of each subject is relatively stable. There is no pupil escape in a normal subject.

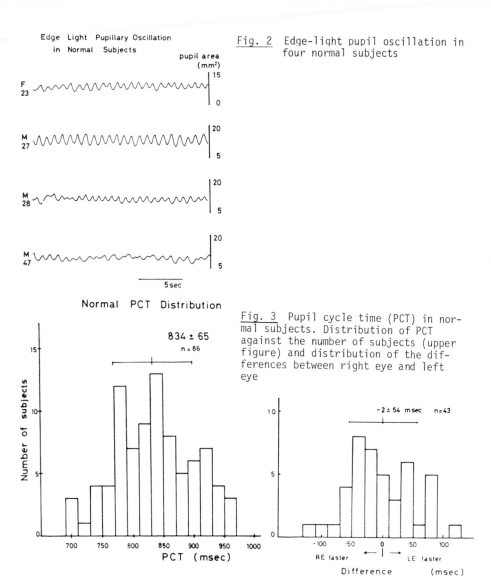

Edge Light Pupillary Oscillation
in Normal Subjects

pupil area
(mm²)

Fig. 2 Edge-light pupil oscillation in four normal subjects

F
23

M
27

M
28

M
47

5 sec

Normal PCT Distribution

834 ± 65

n = 86

Number of subjects

PCT (msec)

Fig. 3 Pupil cycle time (PCT) in normal subjects. Distribution of PCT against the number of subjects (upper figure) and distribution of the differences between right eye and left eye

- 2 ± 54 m sec   n=43

RE faster   LE faster

Difference   (msec)

Figure 3 shows the distribution of the PCT (top) as well as the difference of the PCT between the right and left eyes (bottom) obtained from normal subjects. The distribution of the PCT against the number of subjects is expressed in the magnitude of the histogram. The mean PCT and its standard deviation (SD) were 834 ± 60 msec, which is almost identical with values obtained from previous results among normal subjects [2]. The lower figure indicates the difference of the PCT between the right and left eyes. This value is close to zero; the left eye was 2 msec faster than the right eye. Its standard deviation was 54 msec.

Patients with retrobulbar optic neuritis: In these patients, oscillation was not seen when the vision was extremely reduced. As vision increases, oscillation becomes more noticeable and regular. There were two findings in the patients. First, the oscillation ceased occasionally, and second, there was pupil escape phenomenon. In these cases, the measurement of the PCT was erroneous. As given in Figure 4, the PCTs of every five beats were 881, 932, 884, and 941 msec, respectively. These values are within the normal limit, but, when we calculate the oscillation consecutively as shown in the bottom of the figure, the PCT is 970 msec, which is prolonged slightly. However, when we averaged above 4 numbers, the value was 910 msec. which is not prolonged.

The PCT in patients with retrobulbar neuritis (RBN) was plotted in the ordinate against visual acuity in the abscissa of the affected eye. In addition, the PCTs of the unaffected eye of 15 patients with unilateral RBN as well as normal PCT are shown in Figure 5. The mean value of the PCT of the affected eye among patients with RBN was 851 ± 70 msec and that of the sound eye was 866 ± 58 msec as against a normal value of 834 ± 65 msec. The PCT

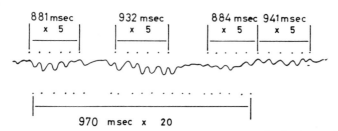

Fig. 4 One example of the method of PCT calculation. Edge-light pupil oscillation and sessions of each part with five beats in the upper number. The PCT of 20 consecutive beats in the lower number

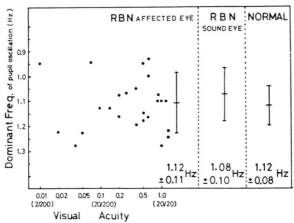

Fig. 5 PCT against visual acuity; left: affected eye by retrobulbar optic neuritis (RBN), middle: sound eye of bilateral RBN, and right: normal control. Vertical bars mean and standard deviation. No prolongation of the PCT is seen.

of both the affected and sound eyes of patients with RBN was slightly prolonged as compared with the normal value. However, there were no statistical
differences among the groups. Even when the visual acuity was reduced, there
were no prolongations of the PCT. When the normal upper limit of the PCT
was set at 950 msec, there were two cases with prolonged PCT in the affected
eye. The PCTs of the affected and the sound eye of the unilateral RBN
patients were compared. In 11 cases, we could measure the PCT bilaterally.
Four cases were excluded because of no pupil oscillation. The mean value
of the PCT in the affected eye was 856 msec, whereas in the sound eye it was
875 msec. Thus PCT of the affected eye was somewhat shorter than that of
the sound eye.

3.2 Computer analysis

Normal subjects: Typical oscillation during a period of 20 seconds and the
power spectrum by Fourier analysis obtained from three controls are shown
in Figure 6. The results of the Fourier analysis are expressed in the magnitude of the histogram. The calibration of the pupil area is given at the
left, and the power spectrum, which is a relative unit of the Fourier analysis, is shown in the right ordinate. The abscissa of the histogram is expressed as frequency in Hz, which is the reciprocal of the PCT. From top
to bottom, the highest peaks are noticeable at 1.10, 1.05, and 1.15 Hz,
respectively (see arrows). When converted PCT, they are 910, 950, and 870
msec, respectively.

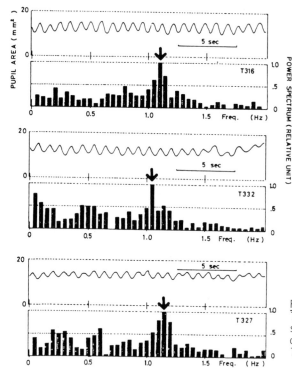

Pupil Oscillation and Power Spectrum
Normal Control

Fig. 6 Three pupil oscillations and their power
spectrum of normal subjects
obtained from Fourier analyses (see arrows)

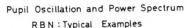

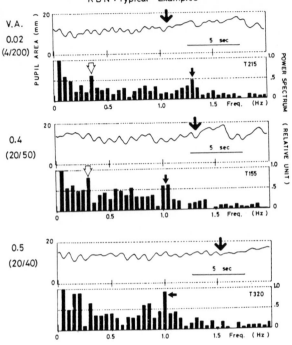

Fig. 7 Typical examples of pupil oscillations with pupil escape phenomenon
(see large arrows), and their power spectrum in three RBN patients (see
small arrows and white arrows)

Affected patients: The results obtained from three patients having different visual acuities are shown in Figure 7. As given by the arrows on the
recordings, pupil wave tended to dilate, which is pupillary escape phenomenon
(see large arrows). The small solid arrows show the frequencies of the major
oscillations, which are from top to bottom 1.40, 1.05, and 1.00 Hz, respectively, or 770, 950, and 1000 msec, respectively. The open arrows indicate
the existence of slow oscillation in two patients at, from the top, 0.35 and
0.30 Hz, respectively, or 2900 and 3300 msec, respectively.

The results obtained from both RBN patients and normal subjects are summarized in Figure 8. In this experiment, the voltage to drive the retinoscope was slightly lower or darker than in previous experiments (analogue
data analysis) because the bulb was changed. The ordinate indicates pupil
oscillation in Hz. Only the dominant frequency in Hz is plotted against the
visual acuity in the abscissa in patients with unilateral RBN of the affected
eye. In addition, those of the sound eyes of patients with unilateral RBN
and of normal subjects were plotted. The mean value and SD of PCT converted
to Hz are 1.12 ± 0.11 in the affected eye of RBN, 1.08 ± 0.10 in the sound
eye of RBN, 1.08 ± 0.08 Hz in normal subjects. Again, even with this detailed analysis, there were no significant differences among the three groups.
In this experiment, pupil escape phenomenon was seen in all cases.

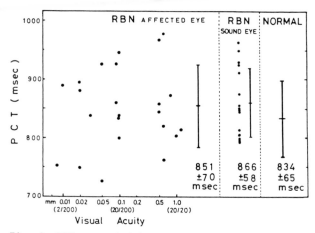

Fig. 8 PCT converted in Hz with visual acuity; left: affected eye (RBN); right: normal control. The ordinate indicates dominant frequency followed by Fourier analysis. No prolongation of the PCT is seen

3.3 Intense Light Stimulation

The pupil in the patients tended to dilate even when an intense light stimulation was given. The results obtained from a normal control and from two patients with RBN are given in Figure 9 (A, B and C respectively). In normal control (A), no exhaustion is seen in the pupil during repetitive light stimulations, and no escape is seen. On the other hand, the patients had a marked exhaustion against time (B and C). This is a typical phenomenon of pupillary escape phenomenon. In each figure, the top recording is the condition of the stimuli and the bottom curve is the pupillary response. Downward deflection is pupillary constriction. The middle curve is the recording of the pupil response. In normal (A), all curves stay flat, whereas in the patients

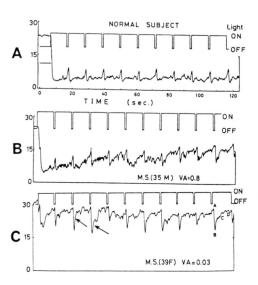

Fig. 9 Pupil response due to intense light stimulation in normal control. No exhaustion can be seen in the pupil (A). Pupil response due to intense light stimulation in patient with RBN (B and C). A marked exhaustion of the pupil against time can be seen in (B), this is a pupillary excape phenomenon and quick exhaustions in each response (see arrows) can be noted in (C), showing another type of pupillary escape phenomenon

(B), the curves start from left bottom and reach to the right up position. As given in (C),advanced patients exhaust more quickly in each stimulus (see oblique arrows). The exhaustion was seen in almost all cases of the examined patients. Therefore, a major finding of the pupil in RBN patients is a pupillary escape phenomenon [5] and is not the prolongation of the pupil cycle time (PCT).

4. Discussion

Present methods of calculating edge-light pupil oscillation seem more accurate than previous techniques utilizing the slit lamp and a stop watch. When the slit lamp microscope is used, on the other hand, it is possible for a minor oscillation to elude measurement. For example, when two examiners calculate the PCT under the slit lamp, frequently two results are obtained from a given patient, one slow and one normal. This is probably due to two reasons: first, when minor oscillations are completely calculated, the PCT would be normal. Second, when some minor oscillations are not calculated and ignored, the PCT would be prolonged. Furthermore, the PCT is affected by various artifacts. Therefore, a more accurate method that would discount the effects of all possible artifacts such as blinking of the eye, face, and body movements is most desirable. This would be particularly true when examining patients with large central scotomas. By using a monitor scope, one is not only able to discount the effects of the above-mentioned artifacts, but one can also detect a slippage of the slit—light stimulus. Fixation of the eye is also controllable by viewing the monitor scope as a slip of the pupil. From the recording chart, one can easily eliminate all artifacts mentioned above.

Wave abnormality, which is the cessation of the pupil oscillation with a dilation of the pupil followed by constriction again was a major oscillating pattern among patients with RBN. More than 80% of the cases showed this abnormality, which was very easy to produce in patients both by edge light pupil oscillation and by intense light stimuli.

In conclusion, wave abnormality is more important in detecting and diagnosing RBN patients. This wave abnormality corresponds to the pupillary escape phenomenon due to afferent pupillary defect, possibly by fatigue of the optic nerve. This will be clinically detected by monitoring the patient pupil under slit lamp microscope with the aid of intense light from the side.

One has to remember that pupil escape sign of very old optic neuropathy becoming simple atrophy with preserved vision less than 20/200 is not so obvious although there is an edge-light pupil oscillation. But you can diagnose such patients from ophthalmological examination with pupillary dilation.

5. Summary

Edge-light pupil oscillation induced by a slit was recorded by an infrared pupillometer. Pupil cycle time was calculated from both recorded charts and from Fourier analyses. There were no statistical differences between patients with retrobulbar optic neuritis and the controls. In the patients, the amplitude of the edge-light pupil oscillations tended to diminish with time as the light was still held on the eye, until it stopped cycling regularly. This pupillary behavior was never observed in normal subjects. The pupil tended to dilate when an intense light stimulus was given. This was considered to be pupil escape phenomenon due to afferent defect and was a unique phenomenon seen in the patients. Therefore, it was concluded that the prolongation of pupil cycle time was not useful for the diagnosis of the patients with optic neuropathy and for that purpose, one should use pupil escape phenomenon.

References

1 H.S. Thompson: Amer. J. Ophthalmol. 62, 860 (1966)
2 S.D. Miller and H.S. Thompson: Amer. J. Ophthalmol. 85, 635 (1978)
3 S. Ishikawa, M. Naito, and K. Inaba: Ophthalmologica 160, 248 (1970)
4 K. Ukai, T. Higashi, and S. Ishikawa: Neuroophthalmol. 1, 33 (1980)
5 W.S. Duke Elder, *Textbook of Ophthalmology*, Vol. 3, 2975, and Vol. 4, 3791 (Mosby, St. Louis, 1941)

Part 6

Clinical Applications
of Visual Psychophysical Testing

Recent Developments in Clinical Contrast Sensitivity Testing

G.B. Arden

Department of Clinical Ophthalmology, Institute of Ophthalmology, Judd Street
London WC1 9QS, England

Since a previous review [2] of contrast sensitivity, there has been a considerable expansion of interest in the value of examining vision using sine wave gratings. It is the aim of this paper to summarize the clinical results obtained, and also the advances in psychophysics and animal physiology which are of importance in understanding the principles of the method. Examination of spatial aspects of vision cannot be absolutely separated from temporal aspects [9,14], but flicker and movement are outside the scope of this review.

When vision is measured with sine wave gratings, various periodicities (spatial frequencies) can be employed, and the threshold contrast determined. The least contrast (less than 1%) is required for coarse gratings of 2-4 c/deg, and the sensitivity to lower or higher frequencies is reduced. The maximum contrast which can be obtained, 100%, is required to see fine gratings of about 30 c/deg. Testing vision with such gratings is equivalent to measuring visual acuity. If lower spatial frequencies are tested, another aspect of vision is explored [2,3,12] and if the contrast sensitivity varies independently at different spatial frequencies, it is possible that abnormalities in contrast sensitivity might be useful in the diagnosis of disease.

That this may be the case is suggested by physiological investigations which have introduced the concept of channels in the visual system which respond to bands of spatial frequencies--with a bandwidth typically of 1.5 octaves [34,28,66]--so that the system which responds best to fine lines and sharp edges may be distinct to that which responds to the gradual changes in contrast seen in coarser gratings. It has been suggested that the lowest spatial frequency channel was centered at 1.5 c/deg, and different mechanisms operated for still coarser gratings [17], which are of particular clinical interest. However, even when much lower spatial frequencies are employed, it is possible to show in adaptation and masking experiments [65,66] that exposure to that frequency reduces sensitivity to it and nearby spatial frequencies. Thus even the coarsest and finest gratings are detected by mechanisms which share similar properties. There is considerable interest in their nature. Coarse gratings can be detected by stroboscopic exposures, and therefore a larger area of retina than the central fovea must be involved. Thus such gratings offer a clinical means of investigating function in regions which do not contribute to visual acuity. The visibility of gratings is determined by the number of repeats (periods) that are exposed. For coarse gratings seen by the peripheral retina, threshold decreases as the number of periods increases [38,39,40,58] till very large retinal areas are involved. It is not possible that single detectors should cover such a large area of the visual field, and threshold is determined by probability summation--i.e. there are a number of similar detectors which occupy adjacent retinal regions. The probability of an output from each detector var-

ies with the intensity of the stimulus. For weak stimuli, the probability
that any one will give an output increases with the size of the stimulus in
predictable ways [40]. Although sensitivity per unit retinal area is always
highest in the region adjacent to the fovea, regardless of the spatial frequency used [50], the decrease of sensitivity as a function of eccentricity
is much steeper for higher spatial frequencies. A rough rule of thumb is
derivable, that the angle on the retina (measured from the fovea) from which
the sensation is derived is the reciprocal of the spatial frequency plus 2°.
Thus it is evident, for many clinical purposes, that the area of interest is
in very low frequencies which are often neglected by physiologists, and which
may be difficult to produce with small displays.

It is of interest to know if there are special detectors of sine waves
and if so where they are located. Retinal receptive fields are circular,
but those of the visual cortex are elongated and thus it is possible that
gratings [19,20,21,34,41,42,43,44] are a specially suitable form of stimulus.
The responses of single cortical cells depend upon whether they belong to
the 'X' or 'Y' system [19,10], but in general demonstrate bandpass characteristics which are similar to those found in psychophysical experiments [66].
It should be noted that there seems to be, within the limits of sampling im-
posed by neurophysiological experiments, a continuous spectrum of center
spatial frequencies for any retinal region. There is no cerebral segrega-
tion of cells which respond selectively to spatial frequency bands, as there
is for orientation selectivity, and this imposes constraints upon the locus
of any condition which affects a narrow band of spatial frequencies: it is
much easier to explain the defect as due to damage to a specific part of the
retina than to a specific part of the cortex. However, if orientation sensitivity is specifically affected, the lesion cannot be subcortical.

The fact that sine waves may be used to investigate the visual system has
suggested that the brain contains harmonic analyzers, and that visual processing consists of a type of piecewise Fourier analysis. It is pertinent
to ask what sort of analyzers would possess the bandpass characteristics of
the hypothetical channels. The main slit-shape of the receptive field of a
typical cortical simple cell is bounded by inhibitory flanks, and beyond
these there may additionally be further regions of excitation. It is possible to predict from such results the responses of the cell to drifting
gratings, and it has been shown that in a good proportion of cases, the receptive field characteristics predict the cell's response to gratings: if
cortical cells perform harmonic analysis, no special cell properties are required [41,42,43,44]. However, there are cells in which the correspondence
is poor, and in any case it depends upon the use of stimuli which are optimal in terms of their drift rate, orientation, contrast and so on. When
more complex or strongly suprathreshold stimuli are used, the visual system
shows pronounced non-linearities.

1. Techniques

The simplicity of the printed sheets of gratings [2] has enabled many workers
to investigate contrast sensitivity, but various electronic displays have
been described, ranging from simple devices assembled from oscillators and
laboratory oscilloscopes [4] to large screen colour displays [7] and special
purpose TV displays run in conjunction with microprocessors [68]. One such
device is made commercially for clinic testing, and incorporates a resident
program and a printer, so that the patient's contrast sensitivity is tested
automatically, and the result printed and tabulated, together with the limits
for the normal.

Electronic devices are of course convenient, and television (electromagnetic) systems have the advantage of large size and high luminance, desirable for low frequency gratings and clinical use. The flicker on ordinary TV is not evident if 100 Hz systems are used, and these can now be obtained as near standard items. Generators for use with standard TV systems can either modify the luminance of the moving spot, within the TV lines, or else luminance can be changed line by line. The latter technique limits screen resolution very severely. In practice, the variation in contrast required is from 0.3% to 60%. This is obtainable, but if luminance is altered by the use of digital techniques, it will be found that an 8-bit digital-to-analog converter does not afford sufficient precision, and it is not yet possible to obtain microcomputer systems which will directly drive D-to-A converters at the required speed.

Non-electronic techniques have been described, involving the use of masks and cylindrical lenses [7,8]. These can be incorporated into standard optical systems, and are therefore useful if other parameters (for example, colour or extremely high luminance) are to be studied. The disadvantage is the difficulty in modifying the display. Gratings may be formed by laser interference fringes and a commercially available instrument has been described [22]. The spatial frequency and contrast are readily altered, and the light source, a red helium-neon laser, is completely safe to view. The disadvantage is that the patient must be carefully positioned in the equipment (which resembles a slit lamp). The advantage claimed is that since the laser system bypasses the optical system of the eye, true neural contrast sensitivity may be measured in the presence of considerable refractive error. However, contrast sensitivity at low spatial frequencies is little affected by even considerable refractive error, no matter how the fringes are produced. The laser system is considerably affected by opacities in the media, which produce glare and reduce contrast [2,3,31,53], so "neural" contrast sensitivity may not be readily obtainable in clinic patients. Paper gratings are available commercially. The relationship of the thresholds measured (in arbitrary units) to contrast is not immediately evident, but such information is not required clinically. The plates are arranged so that the position on the page which corresponds to the normal visual threshold is roughly the same for all spatial frequencies. There is a large change in contrast sensitivity with spatial frequency, so if the crude results are plotted, it may become difficult to tell if the results obtained in any particular case are abnormal. To overcome this problem, some researchers plot the reduction in contrast sensitivity, referred to the normal. Such graphs are often called "visuograms" [53,72]. The printed gratings vary in contrast along the length of each bar, and are uncovered till the patient becomes aware of the presence of the grating on what had appeared a blank page. This psychophysical technique is non-standard, and while it is designed for rapid use introduces a degree of subjectivity into the results. It also increases the variance of the results in the normal population, to varying and unquantified degrees. Various improvements have been proposed: the repetition of the test will obviously reduce variance, or only a small window can be exposed. The plates have been cut into smaller pieces, and incorporated into a forced-choice test, and this is claimed to increase the discrimination of the test very considerably [69]. While such modifications may improve the test, they detract from the simplicity and speed, which are important in clinical use. In many cases, the original simple version of the test will yield an obvious clinical result: if not, and finer discrimination is sought, then electronic devices may be preferred and are essential if spatio-temporal aspects of vision are to be investigated.

2. Variation of Contrast Sensitivity with Age

In infants, preferential looking techniques have established a rapid increase in contrast sensitivity and in performance at higher spatial frequencies during the first two years of life [10,15], and pre-school children tested in a modified maze have nearly adult levels on contrast sensitivity by the age of five [6]. Evoked potential tests have established similar rapid increases in infants by objective means, although there is the additional complication of the maturation of the evoked potential itself [21,24,27]. The study of the amblyopia caused by periods of visual deprivation at known ages leads to the same conclusions, although some plasticity can be seen at more advanced ages [67]. In adults, although systematic changes of contrast sensitivity with age have been reported [60,62] this has not been the case in other series [3,26,52,59] by other workers. This discrepancy is partly explained by differences in technique. Forced choice tests show a lesser loss of sensitivity with age than do those which depend upon a decision about criterion levels [69]. Another factor is that in younger people, acuity but not low-frequency contrast sensitivity may be enhanced [52].

Contrast sensitivity losses in the elderly may be associated with uncorrected presbyopia, unless spatial frequencies of less than 1 c/deg are investigated [2,3]. The onset of cataract, which increases glare, also reduces contrast sensitivity even when visual acuity is not obviously deteriorated [1,31,53]. Factors such as alterations in retinal illumination must also be taken into account. Change in low-frequency contrast sensitivity with age is certainly smaller than the loss of optotype acuity [52]. The onset of other diseases may also affect contrast sensitivity [70], even though the detection of such losses may not be relevant to immediate treatment.

3. Contrast Sensitivity in Specific Disease States

3.1 Amblyopia

Amblyopic eyes demonstrate losses of contrast sensitivity for higher spatial frequencies and in general this corresponds to the loss of visual acuity. Loss of contrast sensitivity for lower spatial frequencies varies depending upon the type of amblyopia. In anisometropic amblyopes, there may be no loss, while in strabismic amblyopes, the loss can be severe [29,30,32,45]. The loss of contrast sensitivity may be confined to one region of the visual field or be generalized [49,61] and this correlates with the type of amblyopia and other clinical features, for example the presence of an area of suppression or eccentric fixation. The measurement of contrast sensitivity with a single sinusoid does not adequately reflect the disability caused by amblyopia [56]. The most profound difficulty is experienced in distinguishing more complex images. An elegant quantification of this is to determine the threshold for distinguishing the phase relationships between two gratings of different spatial frequency which are presented simultaneously [46]. Although contrast sensitivity techniques have been used as a tool to investigate amblyopes, the findings have shed more light on the operation of the normal visual system, rather than assisting in the diagnosis or treatment of the clinical condition.

3.2 Cataract

Early cataracts which cause little reduction in optotype acuity reduce contrast sensitivity [2,36,70,72]. This may cause troublesome symptoms [31] and the condition can be demonstrated (and distinguished from other causes of loss of contrast sensitivity) by the further extreme large loss of contrast sensitivity which occurs if a glare source is introduced near the grating [53].

3.3 Glaucoma

Contrast sensitivity is lost in patients with glaucoma, but the original report [2] compared patients with healthy normals unmatched for age, which exaggerated the effect. In addition, the form of test used (paper gratings with varying contrast) may be associated with increased scores and increased variance in elderly people who suffer from glaucoma, and the effect of miotics on glaucoma patients (or possibly elderly subjects in general) is to decrease contrast sensitivity to a greater degree than in normals. The higher spatial frequencies are also affected by the presbyopia and media opacities more commonly encountered in aging patients [62,64]. Nevertheless, there is a loss of contrast sensitivity [4,5,72] which is specific for the disease [18,69]. It may be increased if the gratings are shown appearing and disappearing [4,5]. It amounts to 6 dB in one series of patients with mild glaucoma--i.e. the threshold is raised fourfold. The loss is not specific to lower spatial frequencies and therefore, in principle, optotype charts could also be used to detect the glaucomatous defect. In practice, however, this cannot be done. The equivalent optotype acuity is 20/25, and charts with this degree of resolution are not available. In addition, the variability of optotype measurements is greater than that of low—frequency gratings, especially in elderly persons, where problems of presbyopia, some forms of cataract and even macular disease increase the variance of the result, so that an acuity of 20/30 or even less must be considered to be within the normal range.

It has been suggested that grating tests could be employed for screening for glaucoma. In a survey of the relatives of known glaucoma, using paper plates of gratings, the test picked up those few undetected cases of frank glaucoma, and it was established that there was a strong correlation between abnormal features in the examination and a high score [33]. The variance on the plates, however, is so great that in the general population the incidence of false positive and negative results is high [64]. This suggests the need for improving testing methods. The use of a forced choice test should be important in this regard [69]. It may also be more important to test contrast sensitivity in the peripheral retina or exclude the macula [49]. It has been reported that a far higher "score" (lower contrast sensitivity) should be used in practical screening, and then, provided there is also loss of visual acuity to some degree, the test is more informative than the use of field screeners, or measurements of intraocular pressure [18].

3.4 Maculopathies

Contrast sensitivity is reduced in maculopathy [2,3,12,25,36], but so is visual acuity. In conditions where the central fovea is involved, it is likely that there is no additional information to be obtained by the use of low spatial frequency gratings. Thus, the provision of Amsler grids (or straight edges) [76] enables patients with disciform degeneration and detachment of the RPE to notice the early onset of distortions of vision in time for laser photocoagulation to be effective in preserving acuity. The exception occurs where the maculopathy is part of a general retinal condition, and parafoveal changes may precede the loss of visual acuity. There have been reports of loss of contrast sensitivity associated with tapetoretinal degenerations [35, 47, 71]. If there is extreme narrowing of the visual field, it is possible that a relatively greater reduction in contrast sensitivity at low spatial frequency could occur. Some patients with retinitis pigmentosa additionally suffer from posterior polar cataracts, which may also explain the reports. It is unlikely that there is a specific loss of low-frequency information in such conditions which would be of diagnostic value: for example, in the case of early cone or central retinal involvement, other clinical methods make the fact very obvious.

Several reports indicate that mild maculopathy can be detected in diabetes, when visual acuity has not suffered [2,3,25,72]. Such findings might be anticipated in view of the patchy nature of the condition. A great deal of variability was found, not surprising in view of the age range of the population investigated, and the likelihood of intercurrent conditions which also affect contrast sensitivity existing in a diabetic population. This reduced the value of the test (printed gratings) in separating the large population which is at risk into groups with no retinopathy, mild background retinopathy, and those with central leakage who require treatment. However, there are no longitudinal studies to indicate whether in each individual the grating scores increase at the stage of the retinopathy where deterioration would warrant medical intervention. The reproducibility and reliability of contrast sensitivity measurements with printed gratings is much greater for repeated tests in one individual than across a whole population [69,70].

In a study of 95 cases of recovery from uniocular contusion, using printed sheets, a reliable relative loss of contrast sensitivity in the affected eye was found in 15 cases [16]. Of these fundal appearances and visual acuity were normal in 7: in two more, the eye which had suffered damage was previously amblyopic, and in the remainder there was visible damage to retina or optic nerve. It seems likely that the injury caused widespread but small areas of damage, which could spare the foveola, but still lead to a reduction in contrast sensitivity to lower spatial frequencies. The patients were able, in this case, to act as their own controls, so the test was sensitive and reliable. In another report it was shown that many young patients with sickle-cell disease have reduced contrast sensitivity [50], although this was not the case in a control group. The losses were unrelated to changes in visual acuity and there was a low correlation with the reduction in contrast sensitivity and the size of the parafoveal avascular zone. It seems probable that the reported loss is related to small areas of avascularity, and of course similar processes of vascular occlusion and new growth occur in diabetic eye disease.

3.5 Optic Neuropathies and Central Diseases

Contrast sensitivity is reduced in most cases of retrobulbar neuritis [2,63, 73,75], a greater proportion of cases demonstrating abnormalities than for evoked potential testing [51]. Considering the ease of the method, it should therefore be more widely employed in neurology. Even in cases of multiple sclerosis without ocular involvement [54,55] there is frequently a loss of contrast sensitivity. This may be in the low or medium frequencies only, so that notches in the "visuogram" occur. Additionally, the loss of contrast sensitivity may be orientation specific. Such findings imply that the abnormalities which are detected are unlikely to be due entirely to demyelinization of the optic nerve. Even were a plaque to be very small and localized, it is unlikely that only those retinal receptive fields destined to form one class of cortical receptive fields could be regularly singled out by the disease process. It is therefore probable that intracortical damage affects the psychophysical test, while the specific delay in the EP measure is due to conduction delays. This means that the psychophysical test could add information to the electrophysiological result, and would account for the increased frequency with which MS may be detected by using gratings. Experimental and

clinical work on methyl mercury poisoning in monkeys [57] and in human popu-
lations reinforces this view. It has been reported that contrast sensitivity
testing is an early and reliable guide to Minemata disease [37]. While the
experiments on monkeys confirm these reports, the clinical test employed
(printed gratings) is often abnormal in hysterical patients or in malingerers,
and in cases of poisoning from the environment such factors must be taken
into account.

In cases of intracranial neoplasms, losses of contrast sensitivity are
seen [11,13] specifically in association with lesions of the occipital and
right parietal regions, and not in the case of lesions of basal ganglia or
more anterior lesions [36]. Contrast sensitivity may also be grossly disturbed in the case of iatrogenic drug intoxications, and in degenerative cerebral conditions. It has been claimed that specific defects in reading are
associated with changes in the ability to detect gratings [48].

4. Screening

The simplicity of the printed gratings has suggested to several groups that
they could be used in screening for disease. The test is acceptable in
general practice [26] and may be more readily accepted than optotype tests
in illiterate rural communities [50]. The results obtained "in the field"
are closely reproduced by repeat testing some weeks later in an ophthalmological clinic [70]. However, in older people, there is a considerable scatter of results [64], which leads to a number of patients giving "false positive" results [64]. In one trial of glaucoma screening only the highest spatial frequencies were tested, which of course are specially susceptible to
losses due to presbyopia, etc. [69]. The information obtainable with such a
test depends upon the cut-off levels. An analysis of results in a glaucoma
clinic [18] suggests that providing the test is carried out on patients with
some loss of visual acuity (and therefore it is accepted that some patients
with glaucoma will give false negative results), contrast sensitivity testing is of greater value than testing the visual fields with a flash-screener.
The problem of patient variability was also troublesome in a series of diabetic patients [25]. However, serial testing of the same patients has not
been attempted.

Although concepts of contrast sensitivity are fairly recent, the principles have been employed for many years. Following publication of a lecture,
I received a letter from a retired ophthalmologist, who inquired how grating
plates differed from the Young test, which he had employed successfully in
the detection of early glaucoma for 40 years. This test [74] consisted of a
book which contained white blotting paper sheets on which were stuck pieces
of blotting papers impregnated with standard inks, forming "a series of spots
of progressively varying degrees of intensity on a white surround. Each spot
has twice the intensity of its successor, the tints varying to one part in
two thousand and forty eight." The light sense was tested by determining if
the patient was able to detect the faintest spot: if that and the next in
the series were not seen, abnormality was certain. Early glaucoma, sympa-
thetic ophthalmitis and albumenuric retinitis of pregnancy were readily detected. Although the language was different, the ideas were exactly the
same, and I had to write back and say that indeed, in all essentials, my
lecture was anticipated 60 years ago.

References

1   R.A. Applegate and R.W. Massof: Amer. J. Optom. Physiol. Optics 52, 840
    (1975)
2   G.B. Arden: Trans. Ophthal. Soc. U.K. 98, 219 (1978a)

3   G.B. Arden: Brit. J. Ophthalmol. 62, 198 (1978b)
4   A. Atkin, I. Bodis-Wollner, and M. Wolkstein: Amer. J. Ophthalmol. 88, 205)1979)
5   A. Atkin, M. Wolkstein, I. Bodis-Wollner, M. Anders, B. Kels, and S.M. Podos: Brit. J. Ophthalmol. 64, 858 (1980)
6   J. Atkinson, J. French, and O. Braddick: Brit. J. Ophthalmol. 65, 525 (1981)
7   M.S. Banks and P. Salapatek: Invest. Ophthalmol. Vis. Sci. 17, 361 (1978)
8   J.L. Barbur and K.H. Ruddock: Biol. Cybernet. 37, 77 (1980)
9   J.L. Barbur and K.H. Ruddock: Biol. Cybernet. 37, 93 (1980)
10   L.D. Beazley, D.J. Illingworth, A. Jahn, and D.V. Greer: Brit. J. Ophthalmol. 64, 863 (1980)
11   I. Bodis-Wollner: Nature 261, 309 (1978)
12   I. Bodis-Wollner: In *Electrophysiology and Psychophysics: Their Use in Ophthalmic Diagnosis*, ed. by S. Sokol (Little, Brown and Company, Boston, 1980)
13   I. Bodis-Wollner and S. Diamond: Brain 99, 695 (1976)
14   I. Bodis-Wollner and C.D. Hendley: J. Physiol. 291, 251 (1979)
15   A. Bradley and R.D. Freeman: Vision Res. 22, 953 (1982)
16   Y. Canavan and D.B. Archer: Brit. J. Ophthalmol. 64, 613 (1980)
17   F.W. Campbell, J.R. Johnstone, and J. Ross: Vision Res. 21, 723 (1981)
18   J. Daub and R. Crick: Ann. Ophthalmol., in press (1983)
19   R.L. De Valois, E.W. Yund, and N. Hepler: Vision Res. 22, 531 (1982)
20   R.L. De Valois, D.G. Albrecht, and L.G. Thorell: Vision Res. 22, 545 (1982)
21   V. Dobson and D.Y. Teller: Vision Res. 18, 1469 (1978)
22   M. Dressler and B. Rassow: Invest. Ophthalmol. Vis. Sci. 21, 737 (1981)
23   J. Enoch, E. Campos, and H. Bedell: Arch. Ophthalmol. 97, 76 (1979)
24   A. Fiorentini, M. Pirchio, and D. Spinelli: Invest. Ophthalmol. Vis. Sci. 19, 950 (1980)
25   J.M. Ghafour, W.S. Foulds, D. Allan, and E. McLure: Brit. J. Ophthalmol. 66, 492 (1982)
26   C.M. Harris: Brit. Med. J. 282, 1279 (1981)
27   L. Harris, J. Atkinson, and O. Braddick: Nature 264, 570 (1976)
28   R.F. Hess: Vision Res. 20, 749 (1980)
29   R.F. Hess and F.W. Campbell: Vision Res. 20, 755 (1980)
30   R.F. Hess, D.C. Burr, and F.W. Campbell: Vision Res. 22, 757 (1982)
31   R.F. Hess and G. Woo: Invest. Ophthalmol. Vis. Sci. 17, 428 (1978)
32   R.F. Hess, T.D. France, and K. Tulaney-Keesey: Exp. Brain Res. 44, 295 (1981)
33   R.A. Hitchings, D.J. Powell, G.B. Arden, and R.M. Carter: Brit. J. Ophthalmol. 65, 565 (1981)
34   R.A. Holub and M. Morton-Gibson: J. Neurophysiol. 46, 1244 (1981)
35   L. Hyvarinen, J. Rovamo, P. Laurinen, and A. Peltomaa: Acta Ophthalmogica 59, 763 (1981)
36   K. Kobayashi: Japan Rev. Clin. Ophthalmol. 74,703 (1980)
37   K. Mikuno, S. Ishikawa, and R. Okamura: Brit. J. Ophthalmol. 65, 284 (1981)
38   J.N. Kroon and G.J. van der Wildt: Vision Res. 20, 253 (1980)
39   J.N. Kroon, J.P. Rijsdik, and G.J. van der Wildt: Vision Res. 20, 235 (1980)
40   J.N. Kroon, J.P. Rijsdijk, and G.J. van der Wildt: Vision Res. 20, 243 (1980)
41   J.J. Kulikowski and P. Bishop: Experientia 37, 100 (1981)
42   J.J. Kulikowski and P.O. Bishop: Vision Res. 22, 191 (1982)
43   J.J. Kulikowski, P.O. Bishop, and H. Kato: Exp. Brain Res. 44, 371 (1981)
44   J.J. Kulikowski and P.O. Bishop: Exp. Brain Res. 44, 386 (1981)
45   H. Ishikawa, S. Sekimoto, and N. Nakano: Acta Soc. Ophthalmol. Japan 83, 1549 (1979)

46 M.L. Lawden, R.F. Hess, and F.W. Campbell: Vision Res. 22, 1005 (1982)
47 C.R. Lindberg, D.A. Fishman, R.J. Anderson, and V. Vasquez: Brit. J. Ophthalmol. 65, 855 (1981)
48 W.J. Lovegrove, A. Bowling, D. Badcock, and M. Blackwood: Science 210, 439 (1980)
49 B.L. Lunch and G. Lennerstrand: Acta Ophthalmogica 59, 21 (1981)
50 R.J. Marsh, S.M. Ford, M.F. Rabb, R.J. Hayes, and G.R. Sergeant: Brit. J. Ophthalmol. 66, 155 (1982)
51 R.B. Matthews, J.R.B. Walton-Bell, and L. Pountney: J. Neurol. Neurosurg. Psych. 45, 303 (1982)
52 C. Owsley and R. Sekuler, in press (1982)
53 L.E. Paulsen and J. Sjostrand: Invest. Ophthalmol. Vis. Sci. 19, 401 (1980)
54 D. Regan, R. Silver, and T.J. Murray: Brain, 100, 563 (1977)
55 D. Regan, J.A. Whitlock, T.J. Murray, and K. Beverley: Invest. Ophthalmol. Vis. Sci. 19, 324 (1980)
56 I. Rentschler, R. Hilz, and H. Brettel: Behav. Brain Res. 1, 433 (1980)
57 D.C. Rice and S.C. Gilbert: Science 216, 759 (1982)
58 J.G. Robson and N. Graham: Vision Res. 21, 409 (1981)
59 R. Sekuler, L.P. Hutman, and C.J. Owsley: Science 209, 4462 (1980)
60 H. Singh, R.L. Cooper, V.A. Alder, G.J. Crawford, A. Terrell, and I.J. Constable: Brit. J. Ophthalmol. 65, 518 (1981)
61 R. Sireteanu and M. Froniua: Vision Res. 21, 1055 (1981)
62 H.W. Skalka: Brit. J. Ophthalmol. 64, 21 (1980a)
63 H.W. Skalka: Brit. J. Ophthalmol. 64, 24 (1980b)
64 S. Sokol, A. Domar, and A. Moskowitz: Invest. Ophthalmol. Vis. Sci. 21, 1509 (1980)
65 C.F. Stromeyer, S. Klein, B.M. Dawson, and L. Spillman: Vision Res. 22, 255 (1981)
66 D.J. Swift and R.A. Smith: Vision Res. 22, 235 (1982)
67 D. Taylor, Vaegan, J.A. Morris, J.E. Rogers, and J. Warland: Trans. Ophthalmol. Soc. U.K. 99, 170 (1979)
68 P.A. Taylor: J. Physiol. 284, 20P (1978)
69 Vaegan and B.L. Halliday: Brit. J. Ophthalmol. 66, 477 (1982)
70 R.G. Weatherhead: Brit. J. Ophthalmol. 64, 591 (1980)
71 M. Wolkstein, A. Atkin, and I. Bodis-Wollner: Amer. J. Ophthalmol. 81, 491 (1976)
72 M. Wolkstein, A. Atkin, and I. Bodis-Wollner: Ophthalmology 87, 1140 (1980)
73 G. Woo and W.F. Long: Austral. J. Optom. 62, 293 (1979)
74 G. Young: Trans. Ophthalmol. Soc. U.K. 38, 279 (1918)
75 R. Zimmern, F. Campbell, and I. Wilkinson: J. Neurol. Neurosurg. Psychiat. 42, 407 (1979)
76 The Macular Degeneration Study Group: Arch. Ophthalmol. 100, 912 (1982)

Contrast Sensitivity Testing in Patients with Juvenile Diabetes

Harold W. Skalka and Harold A. Helms

The Combined Program in Ophthalmology, University of Alabama in
Birmingham, Eye Foundation Hospital, Birmingham, AL 35233, USA

1. Introduction

Grating contrast sensitivity has long been used in the study of visual physiology [1], and more recently in the study of ophthalmological disorders [2,3].
While Snellen acuity tests determine only the highest detectable spatial
frequency at maximal obtainable contrast, grating tests measure the minimum
detectable contrast over a wide range of spatial frequencies, more closely
simulating the visual demands of daily life. Traditionally, contrast sensitivity is defined as the reciprocal of the minimum detectable contrast, and
a plot of contrast sensitivity vs. spatial frequency is a representation of
the contrast sensitivity function (CSF). Many factors are known to influence the CSF, including pupillary diameter and the refractive power of the
eye [1], age of the observer [2], and retinal disease [3]. In this paper
we report measurements of the CSF in a population of 42 juvenile onset diabetics, and 20 age matched non-diabetic controls.

2. Methods

The CSF was measured in 62 naive observers using vertical sinusoidal bar
gratings. Twenty observers (39 eyes) were non-diabetic controls and 42 observers were recruited from the juvenile diabetes service at the Eye Foundation Hospital. All had been admitted for a program of diabetic education
and diet, as well as insulin pump regulation. Pattern blood sugars were
routinely obtained. Diabetic observers were tested near the end of their
hospital stay, were using insulin-infusion pumps, and were known to have
been under good control for at least 24 hours prior to testing.

Prior to testing, all observers underwent a complete ophthalmological
examination. Refractive errors were corrected to 20/20 acuity. If any other
abnormality was noted during this examination, the subject was excluded from
the study.

Testing consisted of presenting each observer with a series of gratings
displayed on a monochrome television monitor (U.S. standard, 525 lines, 60
Hz vertical scan, 2:1 interlace, P4 phosphor). Spatial frequencies of 0.4,
0.8, and 3.2 cycles/degree were used. Mean luminance was always set at 100
cd/m^2, and contrast, defined as $(L_{max}-L_{min})/(L_{max}+L_{min})$, was measured with
a precision photometer before each testing session. Gratings were generated
by an IOL model television signal generator controlled by a Commodore Pet
2001^R microcomputer (the control program and design for the interface cir-
cuit were developed by Arden and associates). Ambient lighting consisted
of a single well-shaded 60W lamp placed ten feet behind the observer. Observers were instructed to occlude one eye with a plastic occluder, to watch
the screen carefully, and to signal the appearance of a grating by pressing

a button. They were informed that the computer was programmed to present the gratings at random and that their responses would be analysed by the computer to detect cheating. (In fact, the computer always started the first trial at zero contrast and each subsequent trial at about 2/3 the contrast level of the previous response, this method of presentation seeming random to all observers tested.) Each observer eye was presented with at least six trials at each spatial frequency, and the mean and standard deviation of his responses were computed and displayed to the examiner in real time. If the standard deviation was greater than 20% of the mean, the observer was assumed to be unreliable and he was excluded from the study. Each eye of every observer was tested sequentially at 0.4, 0.8, and 3.2 c/deg.

Each observer was also tested with the Arden grating test plates. These are high-quality, photographically reproduced gratings of constant frequency which vary continuously from zero to high contrast from top to bottom of each plate. Observers viewed the plates monocularly in a well-lighted area as the examiner slowly uncovered each plate, using a blank card of albedo similar to that of the plate. Observers were instructed to stop the examiner when the grating was first seen, and the highest uncovered number was taken as the score for that plate. Again, both eyes were tested sequentially. Plates displaying spatial frequencies of 0.2, 0.4, 0.8, 1.6, and 3.2 c/deg at the viewing distance were used.

3. Results

Mean scores for each video grating and the cumulative Arden grating test score for each eye were compared to age-matched norms established by SKALKA [4], and the eye was classified as normal or abnormal for each video grating frequency and for the cumulative score on the Arden grating plates. Fig. 1 depicts the percentage of eyes in each group scoring abnormally for each of these four parameters. There was no significant difference between groups for video gratings of 0.4 c/deg. For video gratings of 0.8 and 3.2 c/deg and for cumulative Arden grating scores, significant differences existed be-

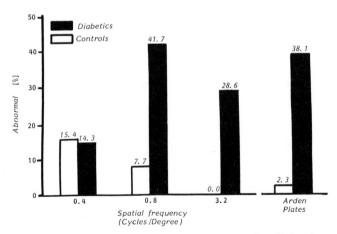

Fig. 1 Comparison of the performance of the diabetic group (shaded bars) and the control group (unshaded bars) for video gratings of 0.4, 0.8, and 3.2 c/deg and for the Arden grating test. The vertical bars represent the percentage of each group scoring abnormally in each category.

tween the groups. At 0.8 c/deg, 41.7% of the diabetic vs. 7.7% of the control eyes scored abnormally (p < 0.001), and at 3.2 c/deg 28.6% of the diabetic vs. none of the control eyes scored abnormally. Finally, the cumulative Arden grating plate scores showed 38.6% of the diabetic vs. 2.3% of the control eyes to score abnormally (p < 0.001).

4. Discussion

We found a significantly increased incidence of abnormal contrast sensitivity in well-regulated juvenile onset diabetics with disease of short duration who were free of retinopathy by routine ophthalmoscopic examination. The deficit in contrast sensitivity among the diabetics seemed to be confined to higher spatial frequencies, but too few frequencies were tested to comment more precisely on differences in the CSF for diabetics vs. non-diabetics. SJOSTRAND and FRIESEN [5] and ARUNDALE [2] reported impaired contrast sensitivity at high spatial frequencies in patients with known diabetic retinopathy. Certainly, many changes must occur in diabetic retinopathy before the disease reaches the point of clinical recognition. Indeed, CUNHA-VAZ et al. [6] reported a marked increase in the permeability of retinal vessels to fluorescein, as measured by vitreous fluorophotometry, in diabetics with normal ophthalmoscopic and fluorescein angiographic findings. Since our results show deficits among our diabetic observers similar to the CSF deficits of patients with known diabetic retinopathy, it is tempting to speculate that the CSF might serve as an early marker for diabetic retinopathy. This must remain only speculation until sufficient time has elapsed for cases of clinical diabetic retinopathy to emerge in our diabetic population, allowing us retrospectively to assess the correlation of early CSF deficits with later development of retinopathy. To this end, we will endeavour to follow as many of our subjects as we can, and are adding new patients to the study.

Both diabetics and controls demonstrated a greater than expected incidence of abnormal scores for the 0.4 c/deg video grating. We believe this represents an effect similar to that reported by HOEKSTRA et al. [7], who found contrast sensitivity to decrease as the number of cycles in the target decreased. Our 0.4 c/deg target contained only 10 cycles. The 0.2 c/deg Arden test plate contains only 5½ cycles, and nearly half the test scores greater than 16 in our diabetic group (18 of 38) occurred with this plate. Whatever the explanation for the 0.4 c/deg results, the same factors were operant in diabetics and controls, allowing comparison between the two groups to remain valid. Future studies will utilize a larger monitor, enabling the display of more cycles of low frequency gratings.

5. Conclusions

We determined spatial contrast sensitivity in a group of juvenile onset diabetics (84 eyes) and a group of age-matched controls (39 eyes), using both sine wave gratings presented on a video monitor and a series of photographic plates. A complete ophthalmological examination, including refractive correction to 20/20 or better, was done prior to testing, and persons with any sign of ocular pathology were excluded from the study. Diabetics were under good control, as documented by pattern blood glucose measurements, for at least 24 hours prior to testing.

Test results were compared to age-matched norms. There was no difference between the two groups for video gratings subtending 0.4 c/deg; but gratings of 0.8 and 3.2 c/deg and the total score on the Arden photographic plates showed a significantly higher incidence of abnormal scores in the diabetic group.

Our results suggest a significantly increased incidence of impaired contrast sensitivity at high spatial frequencies in diabetic patients having no evidence of ocular pathology on routine ophthalmological examination.

References

1 F.W. Campbell and D.G. Green: J. Physiol. 181, 576 (1965)
2 K. Arundale: Brit. J. Ophthal. 62, 213 (1978)
3 M. Wolkstein, A. Atkin, and I. Bodis-Wollner: Ophthalmology 87, 1140 (1980)
4 H.W. Skalka: Brit. J. Ophthal. 64, 21 (1980)
5 J. Sjostrand and L. Frisen: Acta Ophthalmol. 55, 507 (1977)
6 J. Cunha-Vaz, J.R. Faria de Abreu, A.J. Campos, and G.M. Figo: Brit. J. Ophthal. 59, 649 (1975)
7 J. Hoekstra, D.P.J. Van der Goot, G. Van den Brink, and F.A. Bilsen: Vision Res. 14, 365 (1974)

The Effect of CAM Treatment and Occlusion Therapy on Contrast Sensitivity Function in Amblyopia

George C. Woo

School of Optometry, University of Waterloo, Waterloo, Ontario, Canada N2L 5M3

Introduction

Assessment of visual function has traditionally involved only measurement of the limit of resolution. This has been obtained by use of letters or gratings at a specific contrast. This rather restrictive method has limited our appreciation of vision in general. CAMPBELL and GREEN [1] in 1965 used a completely new approach that assessed not only the resolution limit but more importantly the visibility of objects within the resolution limit. These measurements typically involve determining contrast sensitivity over a wide range of grating sizes by means of sinusoidal luminance profiles, thus measuring the contrast sensitivity function. It mirrors the now familiar modulation transfer function measurements of optical and photographic devices.

While WORTH [2] and CHAVASSE [3] are often considered to be the pioneers in the treatment of amblyopia, LYLE and JACKSON [4] report that Buffon in 1743 was the first to attempt this. For over 200 years, the treatment of amblyopia has remained largely unchanged. HESS [5] stated that constant occlusion is known for its variable results and that many months may pass before the maximum achievable acuity is attained. He believed that such an extreme and prolonged treatment approach was psychologically distressing, educationally disruptive, tedious and domestically unsatisfactory. As a consequence few patients follow the treatment plan.

BANKS et al. [6] reported a new treatment for amblyopia which produced a more rapid improvement in visual acuity. The technique required occlusion of the non-amblyopic eye for seven minutes each session. During each session the patient viewed a series of rotating square wave gratings underneath a clear plastic plate on which the patient played games to ensure concentration. In an earlier study, WOO and DALZIEL [7] reported that in the absence of change of visual acuity, contrast sensitivity function can be changed with the use of the CAM vision stimulator.

In this study, four subjects initially underwent the CAM treatment program. Contrast sensitivity functions of four subjects were measured before and after their respective treatment programs in addition to obtaining some pertinent clinical data. A year later, the same subjects underwent conventional occlusion therapy for amblyopia. Contrast thresholds and binocular vision functions were again measured before and after each treatment program. The treatment program required occlusion of the non-amblyopic eye for 4.5 to 5 h per day for a period of approximately 20 days.

Method

Vertical sine wave gratings of varying frequency and contrast were generated on the screen of an oscilloscope (P31 phosphor). The screen's mean luminance was 18 cd/m^2. The contrast was modulated about this luminance and adjusted by the experimenter with a decibel meter. The oscilloscope's immediate surround was illuminated by indirect lighting to approximately the screen's mean luminance. The test distance was varied from 114 cm for spatial frequencies up to 3 c/deg to 570 cm for spatial frequencies higher than 3 c/deg. Angular subtense of the target was either 5° at 114 cm or 1° at 570 cm. Monocular contrast thresholds were determined from non-seeing of the grating target. Each datum point represented the average of five readings.

Each subject underwént a complete eye examination. Good ocular health was a prerequisite for selection and refractive errors were fully corrected. Each subject also underwent a complete binocular coordination examination prior to commencement of the CAM treatment program [6]. The magnitude, direction, and comitance of the ocular deviation was then established. Retinal correspondence, sensory fusion, and the threshold of stereopsis were also determined. The acuity of the amblyopic eye was measured by first using a full letter Snellen chart; second, using a Bailey Lovie log MAR Snellen chart; and third, using a set of interaction-free Snellen letter E's. The state of monocular fixation was investigated using a projection ophthalmoscope and a Haidinger brush apparatus. A complete binocular coordination examination was again performed on each of these four subjects upon completion of the CAM treatment program. A year later, occlusion therapy was instituted. The same procedure described above was repeated before and after the occlusion therapy.

Clinical Data

A.W., a 32-year-old Caucasian female with refractive errors of O.D. +2.75 and O.S. +0.75 x 180 had aided visual acuities of O.D. 6/12 (20/40) and O.S. 6/6 (20/20) on full chart testing and Bailey Lovie log MAR chart and O.D. 6/10/5 (20/35) and O.S. 6/4.25 (20/15) on interaction-free testing. Monocular fixation was steady and centred in both the right and left eyes by projection ophthalmoscopy. She was non-strabismic with low phorias and inadequate fusional reserves. After 15 CAM treatments, the right eye acuity remained unchanged as measured on the full chart, the Bailey Lovie log MAR chart, and under interaction-free conditions. There was no change in her refractive status, monocular fixation, or stereoscopic thresholds. A year later she was asked to participate in the occlusion therapy program. Before and after 20 consecutive days of patching the right eye each day for a period of 4.5 to 5 h, she was given the same assessment. There was no change in visual acuities from the previous year before the occlusion program. After the occlusion program, she reported that she saw better occasionally although there was no change in visual acuities elicited with a full Snellen chart, a Bailey Lovie log MAR chart, or a set of interaction-free letters.

B.B., a 25-year-old Caucasian male with refractive errors of O.D. +0.75-0.25 x 180 and O.S. +2.50-0.50 x 180 had aided visual acuities of O.D. 6/4.5 (20/15) and O.S. 6/9+3 (20/30+3) under full Snellen chart testing and O.D. 6/4.8 (20/16) and O.S. 6/42-3 (20/140-3) with the Bailey Lovie log MAR chart and O.D. 6/4.5 (20/15) and O.S. 6/10.5+2 (20/35+2) under interaction-free test conditions. He exhibited a microtropia on the unilateral cover test and on the 4Δ loose prism test. Also present was a constant left esotropia. After 25 CAM treatments, the left eye's acuity remained unchanged on the full chart and the Bailey Lovie chart, but improved 6/6-1 (20/20-1) on the interaction-free letters.

There was no change in the binocular function findings. A year later, he was asked to undergo the occlusion therapy program. In a thirty day period, his non-amblyopic eye was patched for 4.5 to 5 h per day for 13 days. Patching of the non-amblyopic eye was done consecutively for the 5 days immediately before termination of the occlusion therapy. Visual acuity findings obtained before and after occlusion therapy varied little. They were essentially similar to the visual acuities obtained a year earlier.

J.L., a 22-year-old Oriental female, with refractive errors of O.D. +0.75 and O.S. +4.25-2.25 x 020 had aided visual acuities of O.D. 6/6 (20/20) and O.S.6/60 (20/200) on full Snellen chart testing and on Bailey Lovie log MAR chart, and O.S. 6/4.5 (20/15) and O.S. 6/60 (20/200) on interaction-free testing. She manifested a post-surgical constant left esotropia of 6Δ and left hypertropia of 2Δ on cover tests. No stereopsis was reported, although monocular fixation was steady and centered in the right eye. After 16 seven-minute CAM treatments, the left eye's visual acuity remained unchanged on full Snellen chart testing and Bailey Lovie log MAR chart, but improved to 6/30 (20/100) on interaction-free testing. A year later before participating in the occlusion therapy program, the same clinical data were obtained. After patching the non-amblyopic right eye for 20 consecutive days of 4.5 and 5 h each day there was no change in any clinical findings. Her left eye's visual acuities were 6/60 (20/200) using the regular or the Bailey Lovie chart and 6/30 (20/100) using the interaction-free letters.

E.B., a 12-year-old Caucasian female with refractive errors of O.D. +0.25/-0,25 and O.S. +0.25 had visual acuities of O.D. 6/6 (20/20) and O.S. 6/24 (20/80) on full Snellen chart testing and Bailey Lovie log MAR chart. With the interaction-free test, her acuities were 6/4.5 (20/15) O.D. and 6/15 (20/50) O.S. Monocular fixation, while in the right eye was steady and centred, was nasally eccentric by 4Δ and unsteady in the left eye. She manifested a left esotropia with unharmonious anomalous retinal correspondence. She was aware that she was unable to see well with her left eye. No stereopsis was reported. After 15 CAM treatments, the left eye's acuity improved from 6/15 (20/50) to 6/9 (20/30) with the full Snellen chart and 6/7.5 (20/25) in the left eye. She was prescribed +0.75 OU for close work. A year later, her full Snellen visual acuities were identical to those obtained after the CAM treatment program. After the occlusion therapy, her left eye's visual acuity was 6/6 (20/20) with all three visual acuity tests. Her binocular findings remained relatively the same.

Results

There is very little difference in visual acuity before and after treatment in three out of four subjects with either the CAM treatment program or the occlusion therapy program a year later. The fourth subject E.B. maintained her improvement in visual acuity after the CAM treatment. A year later, occlusion therapy further improved her visual acuity. She was the youngest subject in this study.

Figures 1 through 4 show results of an assessment of the contrast sensitivity function for all four subjects. The standard error of the mean in these figures is equal to or less than the symbol size. All four subjects showed substantial but similar improvement in contrast sensitivity before and after training. It is interesting to note that the non-amblyopic eyes of these subjects also showed an improvement in contrast sensitivity function. The degree of improvement, however, is not uniform across the wide range of spatial frequencies. The changes in contrast thresholds of the non-amblyopic eyes of amblyopes have been shown before [7]. This could be due to the peculiar properties of the non-amblyopic eye of an amblyope [8, 9,10].

213

Figs. 1,2. Captions see opposite page

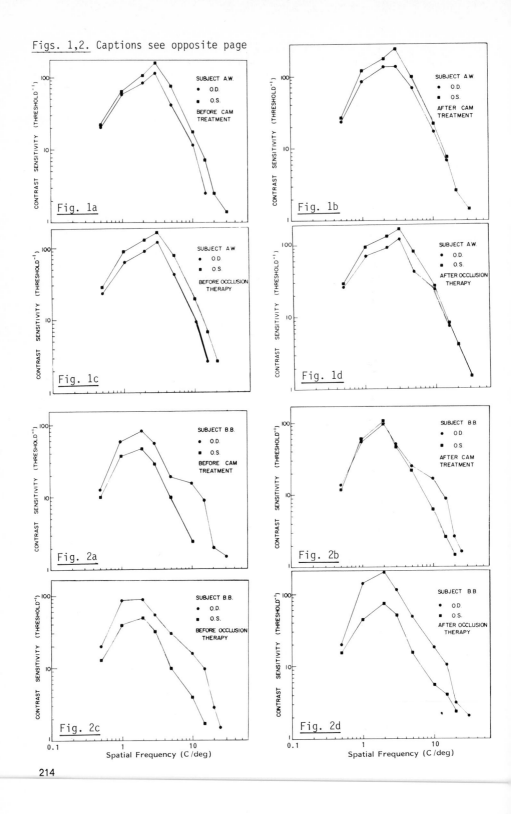

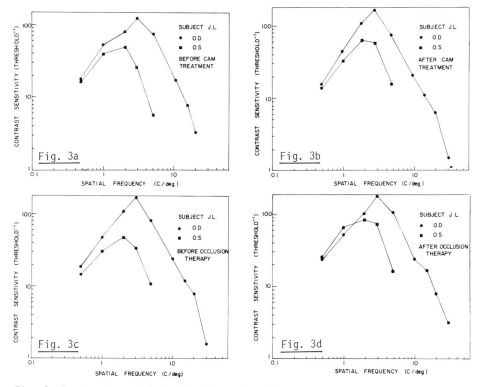

Fig. 3 Contrast sensitivity functions for J.L.

Based upon these results, it can be said that both treatments of ambly-opia improve contrast sensitivity of the amblyopic eye in the absence of improvement in visual acuity. This can be attributed to the fact that the visual improvement of amblyopic patients cannot be demonstrated by tradi-tional measurements of visual acuity. The contrast sensitivity function, on the other hand, provides evidence of an enhancement in their visual func-tions. In all four subjects, the improvements in contrast sensitivity over a wide range of spatial frequencies were not maintained uniformly after a period of approximately 12 months.

In visual training, improvement in visual acuity is often uncertain due to the patient's familiarity with the row of letters presented to him repeat-edly. The speed of reading of these letters also contributes to the degree of improvement. With the use of the CSF technique described in this study, both these problems are eliminated and the quality of vision can be assessed quantitatively.

Fig. 1 Contrast sensitivity functions for A.W.
Fig. 2 Contrast sensitivity functions for B.B.

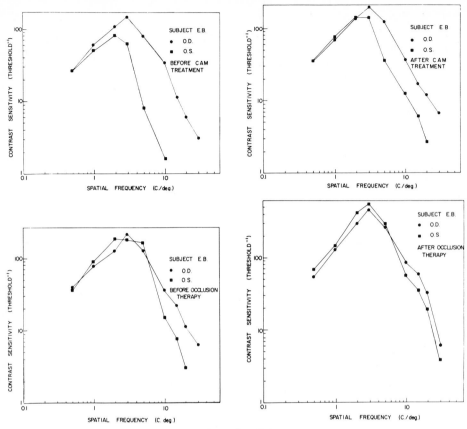

Fig. 4 Contrast sensitivity functions for E.B.

Further investigation is indicated in terms of the length of occlusion
therapy program. What is certain is that there is definitely an improvement
in contrast sensitivity after sessions of 4 to 5 h of occlusion therapy per
day. However, the minimum number of occlusion therapy sessions required before any substantial reduction in contrast thresholds is not known at this
time.

From the results, it appears that age is not a factor in terms of improving the contrast sensitivity function. Improvement can occur in older subjects, and the sensitivity can deteriorate when treatment is stopped.

Conclusions

Contrast sensitivity functions of four amblyopic subjects were obtained in
this study before and after two different amblyopia treatment regimens. In
addition to obtaining contrast sensitivity functions, the acuity of the amblyopic eye was measured by first using a full letter Snellen chart, second
using a Bailey Lovie log MAR chart, and third using a set of interaction-
free Snellen letter E's. The same procedure was repeated after the completion of each treatment program. Both treatments of amblyopia appear to im-

prove contrast sensitivity. However, the improvement was not maintained uniformly in all four subjects. The assessment of the visual function is more complete when the contrast sensitivity function is employed because changes in visual function cannot be precisely determined by means of traditional measurements of visual acuity.

It is concluded that both the conventional method of occlusion for treatment of amblyopia and the CAM treatment program are effective in altering the contrast sensitivity function of amblyopes.

Acknowledgments

I thank Dr. R. Hess of the University of Cambridge for his helpful discussion and Drs. C.C. Dalziel and V. Prentice for collecting some of the clinical data in the first phase of this study. This work was supported in part by a grant from the Natural Sciences and Engineering Research Council of Canada (Number A3449).

References

1   F.W. Campbell and D.G. Green: J. Physiol. 181, 576 (1965)
2   C. Worth: *Squint--Its Causes, Pathology and Treatment* (John Bale Sons and Danielson, London, 1929)
3   F.B. Chavasse: *Worth's Squint*, 7th Ed. (Baillure, Tindall and Cox, London, 1939)
4   T.K. Lyle and S. Jackson: *Practical Orthoptics in the Treatment of Squint*, 3rd Ed. (Blakiston Co., Philadelphia, 1949)
5   R. Hess: Austral. J. Optom. 62, 4 (1979)
6   R.V. Banks, F.W. Campbell, R. Hess, and P.G. Watson: Brit. Orthoptic J. 35, 1 (1978)
7   G.C. Woo and C.C. Dalziel: Acta Ophthalmol. 59, 35 (1981)
8   S.A. Selby and J.M. Woodhouse: Vision Res. 21, 1401 (1981)
9   J.V. Odom, C.S. Hoyt, and E. Marg: Arch. Ophthalmol. 99, 1412 (1981)
10 G.C. Woo: Can. J. Optom. 32, 101 (1970)

A Psychophysical Technique for Testing Explanations of Sensitivity Loss Due to Retinal Disease

Vivienne C. Greenstein, Donald C. Hood, Irwin M. Siegel, and Ronald E. Carr

Department of Psychology, Columbia University, New York, NY, USA and Department of Ophthalmology, New York University Medical Center New York, NY 10016, USA

1. Introduction

Although a retinal disease may affect a patient's ability to differentiate colors and discriminate details, perhaps the most fundamental change is a loss of sensitivity to light. There are many explanations for a disease-related sensitivity loss. For example, a major component of this loss is often attributed to a decrease in quantum catching ability of the photoreceptors. A loss of photopigment, a misalignment of the photoreceptors, or a clouding of the lens or vitreous all decrease the number of quanta caught by the photoreceptors. In this paper, we show how an explanation, like the one above, can be tested using a psychophysical technique. In addition, we reject this particular explanation for the early loss of foveal sensitivity due to retinitis pigmentosa (RP) and suggest instead, that in this disease, and in several other retinal diseases we have studied, sensitivity loss is caused by a decreased responsiveness of individual retinal elements.

In order to specify how a retinal disease affects the responsiveness of the visual system it is best to think in terms of response-intensity functions. A response-intensity function describes the size of the response of the visual system to different intensities of light. The curve in Fig. 1 labeled zero depicts the response-intensity function of a normal visual system. The

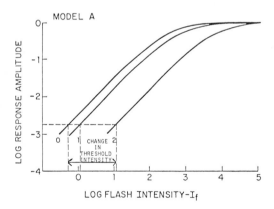

Fig. 1 A hypothetical model for changes in the responsiveness of the visual system. The response-intensity function labeled 0 describes the response of the normal visual system. Symbols 1 and 2 designate different stages of a disease process which causes a decrease in quantum catching ability. For this model, flash intensity is scaled down by a multiplicative constant.

curves labeled 1 and 2 represent different stages of a disease process which
causes a decrease in quantum catching ability. A decrease in quantum catching ability is equivalent to decreasing the effective intensity of a stimulus by a multiplicative constant. That is, if only one half of the quanta
are absorbed compared to the normal, then all stimulus intensities will be
one half as intense. Another way to explain this is that "normal" responses
can be produced in the abnormal system if the stimulus intensity is doubled.
(On a log intensity axis, as in Fig. 1, the curves are shifted horizontally.)
To relate the traditional measure of threshold to these response-intensity
functions, consider the dashed lines in Fig. 1. It is clear that in progressive retinal disease an increase in the threshold intensity is needed
to obtain a small criterion response. Notice that by displaying the loss
of sensitivity in the form of response-intensity functions, it is immediately clear that this explanation makes predictions about both threshold
and suprathreshold responsiveness. Clearly a measure of suprathreshold responsiveness is needed.

2. Methods

2.1 The Probe-flash Paradigm

We have used a psychophysical procedure called the probe-flash paradigm to
assess suprathreshold responsiveness (see Fig. 2A). The observer is required
to detect the presence of a small, brief light, the probe, presented upon a
second light, the 500 msec flash. There is also a large field, the adapting
field, which is continuously present to control the adapting state. Some
sample data are presented in Fig. 2B. These data are increment threshold
data, but unlike the usual increment threshold data, they have been obtained
on flashed backgrounds. Increment threshold data obtained on steady backgrounds would fall on a slope of 1.0 as predicted by Weber's Law. Notice
that the probe-flash data deviate markedly from this slope.

2.2 The Effects of Retinal Disease on Probe-flash Data

Let us suppose that the disease process changed only the quantum catching
ability of the photoreceptors and that the photoreceptors were normal in
every other aspect. (The type of change produced in the response-intensity

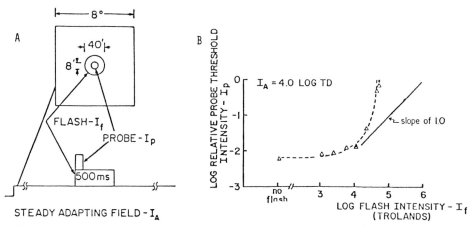

Fig. 2A Spatial and temporal paradigm (see text for details)
Fig. 2B Increment threshold results obtained on flashed backgrounds.

functions is shown in Fig. 1.) According to this explanation for sensitivity loss, the effective intensity of a stimulus has been decreased by a constant. In terms of probe-flash data plotted on a log-log plot, the predicted curve would be shifted up and over to the right by approximately equal amounts compared to the normal (see Fig. 3).

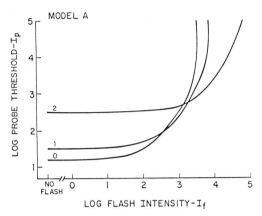

Fig. 3 Hypothetical probe-flash curves derived from the model in Fig. 1 (see Hood and Greenstein, 1982 for details)

To illustrate the advantages of an approach that uses the probe-flash procedure to test alternative explanations for sensitivity loss, we will consider data from our recent study of early retinitis pigmentosa (RP). RP is an example of a disease in which sensitivity loss is frequently said to be due to a decrease in quantum catching ability of functioning photoreceptors. For example, histopathological studies of eyes with advanced or moderately advanced RP have demonstrated that the photoreceptors in peripheral and foveal retinal areas had outer segments which were shortened, twisted, and/or distorted [1,2,3]. Twisting or misalignment of the photoreceptors would result in a reduction in the effective intensity of the stimulus. Also shortening of the outer segments would result in a decrease in the amount of photopigment and therefore in a decrease in quantum catching ability. Densitometry measurements have provided evidence for reduced rod photopigment content [4, 5], and the results of recent psychophysical studies have been interpreted in terms of reduced cone pigment density [6,7]. On the other hand there is some evidence that early cone system changes may not be due to a reduced amount of photopigment [8]. In our study of early RP, we tested the decreased quantum catching hypothesis for foveal sensitivity loss.

The probe-flash data we collected on six patients with early RP are shown in Figs. 4A and 4B. The lower solid curve is the best fitting curve found by eye to fit the median probe-threshold data for five unaffected observers [9]. For the patients, it is clear that probe thresholds are increased for all flash intensities compared with the curve for the unaffected observers. The dashed curve represents the predicted change in shape of the probe-flash curve if sensitivity loss was due to a decrease in quantum catching ability. This predicted curve does not describe the data. However the probe-flash data can be fit reasonably well by a vertical shift of the normal curve (see Fig. caption for details).

220

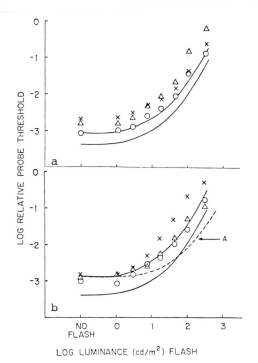

Figs. 4A and 4B Probe-flash data for six patients with retinitis pigmentosa (data for three patients in Fig. 4A, o, Δ, and x, and three in Fig. 4B). The lower curve is the best fitting curve found by eye to fit the median probe-flash data for five unaffected observers. It has been shifted up to fit the data obtained on one patient in Fig. 4A and one in Fig. 4B. The dashed curve A represents the predicted change in the probe-flash curve if all sensitivity loss was due to a decrease in quantum catching ability

Before discussing possible explanations for these results, it is interesting to note that we found similar changes when patients with diabetic retinopathy or macular degeneration were tested [10]. Figures 5 and 6 show typical probe-flash data collected on four patients with diabetic retinopathy, and four with senile macular degeneration. As in the case of early RP, probe threshold is increased for all flash intensities. The open circle in each panel represents the median threshold to the probe alone for seven unaffected observers. The lower smooth curve is the best fitting curve for unaffected observers. An upward displacement of this curve fits the data collected on both groups of patients reasonably well.

It appears that different retinal diseases have produced similar changes in probe-flash data. Let us consider what these data imply about the response intensity function of the diseased visual system. The data in Figs. 4, 5 and 6 are consistent with a response intensity function for the disease state that exhibits a decreased responsiveness to all light intensities [11]. In other words, no matter how intense the light, the amplitude of the response in the abnormal system will always be less than in the normal. A possible change in the response function is shown in Fig. 7. In this case the response size is scaled down by a multiplicative constant (see figure caption for other possible changes). Response scaling like this could result if some of the retinal elements were missing. It could also be due to a decreased responsiveness to light of individual elements. In the case of early RP or diabetic retinopathy, it is unlikely that these diseases could reduce the receptor population in the foveal region sufficiently to produce the changes we see. According to the simplest model a change in sensitivity of 0.3 log unit would require a loss of function of half the receptors. Al-

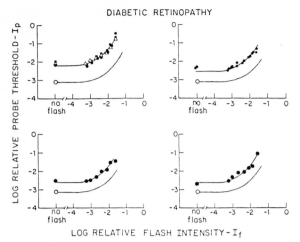

Fig. 5 Individual probe-flash functions obtained from 4 patients with diabetic retinopathy. Repeat measurements were obtained on 2 patients, 2 weeks (filled squares) and 3 months (open triangles) after the initial examination. The open circle in each panel represents the median threshold to the probe alone for the control group (n = 7).

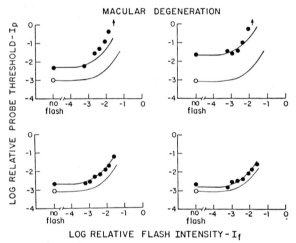

Fig. 6 Individual probe-flash functions obtained from 4 patients with macular degeneration.

though the loss of functioning receptors may contribute to the sensitivity loss we measured, it is unlikely to be the primary cause. A more reasonable explanation is that there is a decreased responsiveness to light of individual elements. A number of factors such as anoxia, decreased metabolic activity, or a change in ionic environment could cause changes in membrane potential and contribute to a decreased responsiveness of individual elements. There is no question that hypoxia plays a role in the pathogenesis of diabetic retinopathy, and it is possible that factors similar to the ones listed

222

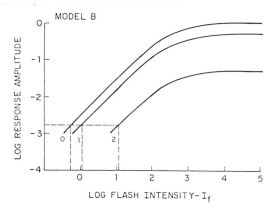

Fig. 7 A hypothetical model for changes in the responsiveness of the visual system. For this model the size of the response to all flash intensities is scaled down by a multiplicative constant. Other possible changes in the response function consistent with the probe-flash data include different amounts of response scaling at high versus low intensities.

above are contributing to the foveal sensitivity loss in early RP. With regard to the probe-flash data obtained on four subjects with macular degeneration, a number of factors, perhaps acting via changes in membrane potential could be implicated. (In addition, fewer retinal elements due to atrophy may contribute to the sensitivity loss.)

3. Summary

In conclusion, there are many explanations for sensitivity loss caused by retinal disease. Provided these explanations can be phrased in terms of response-intensity functions a psychophysical paradigm, the probe-flash paradigm can be used to test among these alternatives.

We have seen that a loss of pigment of a misalignment of receptors does not seem to contribute to foveal sensitivity loss in early RP, diabetic retinopathy, or macular degeneration. In fact, in all these diseases a general decrease in responsiveness of retinal elements may be the major factor leading to a loss of visual sensitivity. Indeed, this may well be a general characteristic of most diseases of the retina.

References

1   H. Kolb and P. Gouras: Invest. Ophthalmol. Vis. Sci. 13, 387 (1974)
2   R.B. Szamier and E.L. Berson: Invest Ophthalmol. Vis. Sci. 16, 947 (1977)
3   R.B. Szamier, E.L. Berson, R. Klein, and S. Meyers: Invest. Ophthalmol. Vis. Sci. 18, 145 (1979)
4   V.N. Highman and R.A. Weale: Am. J. Ophthalmol. 75, 822 (1973)
5   H. Ripps, K.P. Brin, and R.A. Weale: Invest. Ophthalmol. Vis. Sci. 17, 735 (1978)
6   R.S.L. Young and G.A. Fishman: Invest. Ophthalmol. Vis. Sci. 19, 967 (1980)
7   R.S.L. Young and G.A. Fishman: Vision Res. 22, 163 (1982)
8   M.A. Sandberg, M.H. Effron, and E.L. Berson: Invest. Ophthalmol. Vis. Sci. 17, 1096 (1978)

9   V.C. Greenstein, D.C. Hood, I.M. Siegel, and R.E. Carr: Retinitis pig-
    mentosa: A psychophysical test of explanations for early foveal sensitiv-
    ity loss. Submitted for publication (1983)
10 V.C. Greenstein, D.C. Hood, and C.J. Campbell: Invest. Ophthalmol. Vis.
    Sci. 23, 102 (1982)
11 D.C. Hood and V.C. Greenstein: Invest. Ophthalmol. Vis. Sci. 23, 96
    (1982)

Dynamics of Visual Adaptation are Altered in Vascular Disease

G. Haegerstrom-Portnoy, A.J. Adams, B. Brown, and A. Jampolsky

Smith-Kettlewell Institute of Visual Sciences, San Francisco, CA, USA and
School of Optometry, University of California, Berkeley, CA 94720, USA

Introduction

The eye has a relatively fixed contrast sensitivity over a large range of
normal daytime light levels. However, sudden changes in the environmental
light level require that the eye readjusts in order to achieve the same contrast sensitivity at the new level, a process which takes many seconds or
minutes when the new environment is considerably dimmer than the previous
level. During this recovery time, the eye is relatively blind to fine detail, its contrast sensitivity being reduced while it adapts. The initial
stages of recovery are rapid and generally thought to involve neural processes in the retina [1], whereas the predominant slow phase is related to
the rate of regeneration of the bleached photopigment. Any drug or disease
that affects the photoreceptors, the pigment epithelium, or the adjacent
choroid is expected to prolong the recovery of sensitivity because the recovery is dependent on regeneration of the photopigment, which is located in
the outer segments of the receptors. In fact, certain drugs such as alcohol,
even in low doses [2], and antimalarial drugs [3] have been shown to prolong
recovery of sensitivity. It has also been demonstrated that ocular diseases
such as senile macular degeneration, retinal pigment epithelial detachment,
and central serous retinopathy significantly delay recovery even when visual
acuity is normal or only marginally affected [4,5,6,7,8]. Processes that
interfere with the active metabolism required to perform the regeneration of
visual pigments such as decreased retinal vascular supply or from the chorio-
capillaries are also expected to prolong recovery.

The vascular status of the eye is a reflection of the vascular status in
other parts of the body, such as the brain and the kidneys. Consequently
this study was initiated in order to determine whether a test of recovery of
visual sensitivity after light exposure could be used as an indicator of
general vascular status. The experiments involved patients with known systemic vascular disorders (diabetics and hypertensives) and patients who have
been reported to be at risk for vascular complications (women on oral contraceptives). The results of these patient groups were compared to a group of
normal subjects.

Methods

All measurements were made using a glare recovery instrument developed by
Optitec (San Francisco) and based on their patent. Before the beginning of
the test, the patients pre-adapted for at least 5 minutes to the low photopic
luminance levels of the laboratory. To begin the test, each patient wearing
optimal refractive correction looked into an instrument where a circular 5
min of arc flashing target was presented on a background of 23 cd/m^2 at optical infinity. The patient adjusted the luminance of the target until he

was just able to detect it. Three measurements of this contrast threshold were taken and averaged before each glare exposure. The patient then fixated the center of the 20° glare field which had a luminance of 5.6×10^4 cd/m^2. Immediately after a 10 second exposure to the high intensity field the patient's gaze was directed to the center of the test stimulus configuration (Fig. 1) where the 5 min of arc target, set at a preset contrast, was intermittently presented (125 ms flashes at 4 Hz). When the patient recovered sufficient sensitivity to detect the flashing spot, he pushed a button which reduced the contrast of the small spot to another, lower, preset contrast level below the patient's threshold. When sensitivity had recovered further and the spot was detected again, the patient pushed the button again. The times taken to recover to five pre-set contrast levels were recorded. The contrast levels were chosen to give approximately equal time periods between recorded points. Contrast of the target was always the independent variable and is defined as (target − background luminance) divided by (target + background luminance). Log target contrast for the five levels were: 1.89, 1.65, 1.55, 1.44, and 1.28. Each eye was measured twice with a 5 min. recovery period between measurements.

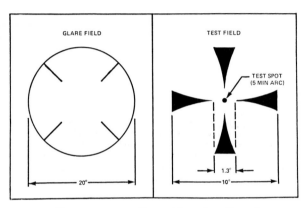

Fig. 1   The target configuration for the glare recovery test

In addition, all the normal, healthy subjects returned after a few weeks and the test sequence was repeated in order to establish the reliability of measures. Visual acuity measurements and funduscopic results were also recorded at each visit.

Analysis

For each patient and for each glare recovery measure, we determined the time constant of the exponential recovery curve. This was found by plotting recovery time (sec) against log target contrast using a procedure that takes the contrast threshold measurements before each glare exposure into account. The procedure involves fitting an equation of the kind

$$y = ae^{bx} + c$$

to the data where the x values are the recorded recovery times and the y values represent the contrast levels of the target. Least squares linear regression analysis was performed by taking the natural logarithm of both sides of the equation

$$\ln (y-c) = \ln a + bx$$

and then determining the slope and the intercept. The inverse of the slope "b" is the time constant and the constant "c" represents the contrast threshold. The time constant is the time taken to recover 68% of the full amplitude. The mean and standard deviation of both the time constant and the contrast threshold were also calculated for each group.

Results

1. Normal Healthy Group

The importance of correcting for individual differences in the ability to detect the target in the absence of glare is demonstrated in Fig. 2.

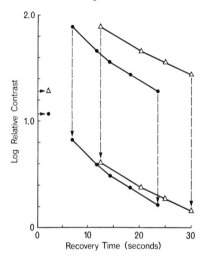

Fig. 2 Glare recovery curves are shown for two patients before and after adjustment for individual differences in contrast threshold. Each patient's contrast threshold before glare is indicated by arrows on the ordinate

Recovery curves after a 10 second high-intensity exposure are shown for two patients in the upper portion of the figure. One of the patients (solid circles) recovers quickly and detects the test target at the first pre-set contrast level after approximately 7 seconds and requires about 12 seconds to see the target at the second contrast level. The other patient (open triangles) requires 12 sec and 20 sec respectively to recover to the same contrast levels and is unable to see the target at the lowest contrast level. The large difference in their response is deceptive however, because the patient with the slower response has a higher contrast threshold than the patient who recovers faster. The contrast thresholds for the two are indicated by arrows on the ordinate. Elevated contrast thresholds in the absence of glare may be caused by differences in pre-retinal media absorption, uncorrected refractive errors or abnormal retinal contrast detecting mechanisms. We are interested in the dynamics of the retinal response and can factor out the effect of contrast threshold on recovery time by subtracting each patient's contrast threshold from the pre-set target contrast thresholds. This is indicated by the arrows shifting each patient's curve down by an amount equal to his threshold before glare (Fig. 2). Now the responses of the two patients to the same "effective contrast" levels are seen to be quite similar. In fact, the analysis shows that the time constants for these two patients are quite similar to each other (16 sec vs. 17 sec).

227

Whenever a new test is devised, the normal response range must first be established. We determined the norms for the instrument by testing thirty healthy adults (between the ages of 18 and 34, mean 24.4). They all had visual acuity better than 20/25, no ocular or systemic abnormalities or history of chronic disease as determined by examination by a physician. Each subject in the normal group was tested a second time between 3 and 6 weeks after the first measurement. The mean time constant was 13.4 sec. The results for the right eyes of the normal group are shown in Table 1.

Table 1

	Time Constant		Contrast Threshold	
	Mean (Sec)	S.D.	Mean	S.D.
Normal Group (N=30) Mean age 24.4 yrs (18-24 yrs)	13.4	6.25	1.15	0.14
Vascular Disease Group (N=12) Mean age 50 yrs (31-63 yrs)	20.1**	7.26	1.33*	0.14
Diabetic Group (N-10) Mean age 35.5 yrs (19-52 yrs)	27.1*	11.98	1.28*	0.14
Hypertensive Group (N=10) Mean age 43 yrs (30-55 yrs)	25.0*	11.14	1.12	0.17
Oral Contraceptive Group (N=12) Mean age 33.3 yrs (30-42 yrs)	25.5*	11.60	1.04	0.24

Comparison with normal group: *p < .01; **p < .05 t test for unequal N.

In order to determine the reliability of the test, comparisons were made between the results from right and left eyes and between the results of the first and second visit. The interpolated glare recovery time for a fixed target contrast was determined using the best fitting equation for the glare recovery measures. The reliability of the test was then assessed by calculating correlation coefficients; the correlation between the results from right and left eyes was 0.82 and the correlation coefficient for the results of the first visit versus the second visit was 0.80. The test shows both reliability for between eyes measures and across time.

2. Systemic Vascular Disorders

Having established the reliability of the glare recovery test, we were interested in how sensitive it would be in patients with compromised vascular supply. Twelve patients between the ages of 31 and 63 years (mean 50.0 years) participated. All patients had visual acuity of 20/25 or better in at least one eye. All of the patients were diabetic requiring insulin and/or had systemic hypertension. Only patients with no or mild retinopathy were included.

In addition, most of the patients had other risk factors for vascular events, such as heavy cigarette smoking, obesity, hyperlipoproteinemia or history of eclampsia or pre-eclampsia.

Even though glare recovery was measured in each eye, only the results from the eye with the better acuity was included in the analysis. The mean time constant of recovery for this group was 20.15 sec, significantly longer than the time constant for the normal group ($p < .05$, t test for unequal N). In addition, this group of patients showed elevated contrast thresholds in the absence of glare which were significantly different from the normal group ($p < .01$). Increased scatter in the media due to the large difference in mean age between the groups (50 vs. 24.4 years) is a possible explanation for the difference. It is also possible that static retinal contrast-detecting mechanisms are abnormal in this group of patients in addition to their abnormal kinetics.

3. Diabetic Group

The glare recovery times for the vascular disease group were abnormal but because this group had both diabetes and/or hypertension as well as diabetic and/or hypertensive retinopathy to various degrees, it is not clear whether these results were caused by the diabetes, the hypertension, the combination of diabetes and hypertension representing general vascular status, or the retinopathy. We therefore decided to test a group of diabetics (N=10) with no other complications such as obesity or hypertension as well as a group of patients with only hypertension (N=10). Half of the patients in each group were free of any ophthalmoscopically visible retinopathy.

Five insulin-dependent diabetics without retinopathy between the ages of 19 and 52 (mean 36 years) participated. They all had long-standing diabetes; the mean duration was 14 years. The five insulin dependent diabetics with mild to moderate background retinopathy were between the ages of 29 and 43 (mean 35 years) and had also been diabetic for a long time (mean 17 years). Each eye was tested, but only the eye with the best acuity was included in the analysis. There were no significant differences between the results from the patients with retinopathy or without retinopathy so the results were collapsed across groups and shown as the "diabetic" group in Table 1. We attach no particular significance to this lack of difference between diabetics with and without retinopathy because of the samll sample size. The mean time constant for the group was 27.1 sec, significantly longer than for the normals. The diabetics also had abnormal contrast thresholds which again could be attributed to either pre-retinal or retinal changes. The diabetic group is only 10 years older than the normal group and we would not expect large changes in the clarity of the media. In any event, the kinetics of adaptation for the group are severely abnormal even when differences in contrast threshold have been accounted for. The abnormality must exist beyond the ocular media, probably in the retina.

4. Hypertensive Group

The hypertensive group consisted of 10 patients between the ages of 30 and 55 years (mean 43 years). They had been treated for hypertension for an average of 12 years. We only included the results from the eye with the better visual acuity, which was 20/25 or better in all cases. No statistically significant differences were found between the 5 patients without retinopathy and the 5 patients who had moderate hypertensive retinopathy. The results for the combined group are shown in Table 1. The dynamics of adaptation were clearly abnormal in the hypertensive group. The time constant was 25.0 sec compared to 13.45 sec for the normals. This difference in time constant is statistically significant at the 0.01 level. No differences from normal in contrast threshold were observed for the hypertensive group in spite of a 20 year difference in age.

5. Oral Contraceptive Group

All three groups with vascular disease showed prolonged recovery times and we were curious to find out whether women taking oral contraceptives might also have prolonged recovery times, since there have been many reports of increased risk for vascular complications in women who take oral contraceptives [e.g. 9,10]. Twelve women between the ages of 30 and 42 years participated (mean 33.3 years). They had been taking oral contraceptives for an average of 5 years. They were a reasonably heterogeneous group in terms of whether they were presently using oral contraceptives and in terms of the length of time since they discontinued the medication. They all had visual acuity better than 20/25 and no known ocular or systemic diseases. The mean time constant for the group was 25.5 sec, considerably longer than the normal group; (p < .01, t test for unequal N). They had no trouble detecting the test target prior to glare; in fact, their contrast thresholds were lower than for the normal group, indicating slightly superior contrast sensitivity. Again, it appears that the dynamics of adaptation of the retina are sensitive to subclinical changes in retinal function.

Discussion

Glare recovery or photostress had been used clinically in the past to test macular function in disease and has been shown to be prolonged in a multitude of diseases including senile macular degeneration, central serous retinopathy and retinal pigment epithelial detachment [4,5,6,7,8]. The present study initially established normal values for a glare recovery test with rigorously defined parameters and measured multiple recovery times from a single exposure. We computed a time constant of recovery which is independent of contrast sensitivity differences between individuals and consequently free of pre-retinal media contamination. The reliability of the test was assessed in a group of young healthy adults and was found to be very high (N=30, r= 0.80). The test was performed on several groups of patients with known vascular disease and a group of young women who were using or had used oral contraceptives and might be considered at risk for vascular events.

In all groups, the mean glare recovery time was significantly prolonged even after differences in contrast threshold had been accounted for. In order to assess how many individuals in each group had altered dynamics of adaptation, we calculated a predicted recovery time to a fixed contrast level using each person's derived equation. Seventy percent of the patients in the vascular disease group showed recovery times more than one standard deviation longer than the mean for the normals. In the hypertensive group, 60% of the patients had longer recovery times using this criterion, while 90% of the diabetic group fell into this category. The equivalent proportion for the women who had been taking oral contraceptives was 75%.

The mechanism of action in the various groups may or may not be the same even though a common underlying factor for all groups is probably general vascular alteration. There is a consensus that one of the underlying causes of diabetic retinopathy is hypoxia caused by abnormal capillary blood flow (for a review see LITTLE [11]). The retina is particularly susceptible to hypoxia because of its high metabolic activity and because of its unusual vascular anatomy. Hypoxia induced in normal subjects has been shown to reduce the rate of adaptation [12] and has also been shown to depress brightness sensitivity at a steady level of adaptation [13]. Anoxia may be a possible mechanism for the prolonged glare recovery times in patients with diameter, since both contrast sensitivity and rate of recovery are depressed by reduced oxygen levels.

A similar argument can be made to explain the results for the patients with hypertension. Hypertension is a disease in which there is an increased resistance to blood flow caused by an unexplained constriction of the arterial beds throughout the body which in turn results in anoxia [14].

The mechanism responsible for the changes we found in the oral contraceptive group is harder to elucidate. It is possible that changes in blood coagulation parameters, rheology and vessel anatomy shown to occur in women taking oral contraceptives may disturb the retinal blood supply and lead to changes in glare recovery [15].

Conclusion

It is clear that tests of glare recovery can be extremely sensitive to changes in retinal integrity. The results on the patients in this study provide evidence for functional loss prior to visual acuity loss or visible fundus change in four quite different groups who at least have in common systemic vascular disturbance or are in a position of increased risk for vascular events. Furthermore we have identified techniques for separating out confounding preretinal changes from retinal changes; the kinetics of glare recovery almost certainly lie at the retinal level. Such tests should provide clinicians with sensitive indices at the early stages of both primary retinal disorders and secondary ocular changes produced by systemic disease.

References

1   H.S. Baker: J. Opt. Soc. Am. 53, 98 (1963)
2   A.J. Adams and B. Brown: Nature 257, 481 (1975)
3   R. Carr, P. Henkind, N. Rothfield, and I. Siegel: Am. J. Ophthal. 66, 738 (1968)
4   S.L. Severin, J.Y. Harper, and J.F. Culver: Arch. Ophthal. 70, 593 (1963)
5   S.L. Severin, R.L. Tour, and R.H. Kershaw: Arch. Ophthal. 77, 163 (1967)
6   J. Glaser, P.J. Savino, K.D. Sumers, S. McDonald, and R.W. Knighton: Am. J. Ophthal. 83, 255 (1977)
7   G. Chilaris: Am. J. Ophthal. 53, 311 (1962)
8   H. Magder: Am. J. Ophthal. 51, 147 (1960)
9   R.I. Hardin: Prog. Cardiovasc. Dis. 15(4), 395 (1974)
10 Collaborative Group for the Study of Stroke in Young Women: JAMA 231, 718 (1975)
11 H. Little: Trans. Am. Ophth. Soc. 74, 573 (1976)
12 J.L. Kobrick and B. Appleton: J. App. Physiol. 31, 357 (1971)
13 S. Hecht, C.D. Hendley, S. Frank, and C. Haig: J. Gen. Physiol. 29, 335 (1946)
14 S. Duke-Elder (ed.): *System of Ophthalmology*, Vol. 10: *Diseases of the Retina* (Henry Kimpton, London, 1967)
15 M. Dugdale and A.T. Masi: J. Chronic Dis. 23, 775 (1971)

Influence of Variable-Sized Backgrounds
on a Hyperacuity Threshold

Rick A. Williams, Edward A. Essock, and Jay M. Enoch

School of Optometry, University of California, Berkeley, CA 94720, USA

Introduction

Previous work [1] has established the clinical utility of assessing the increment threshold of a test spot centered on a variable-sized background of light. The presence of light near the increment threshold test spot systematically changes the observer's increment threshold. With increasing background diameters, thresholds first rise, and then fall to asymptotic levels [2,3], thereby effectively mapping out light-produced spatial interactions about the increment threshold target. The clinical importance of this "desensitization-sensitization" effect is that the loss of one branch of the standard pattern of spatial interactions is indicative of inner retinal pathology at the retinal test point.

We are now extending our clinical testing capabilities to more central levels within the visual pathway. We utilize tests of hyperacuity (e.g., vernier acuity) in this work. That even a novice observer can reliably align two stimulus features to within 3 or 4 seconds of arc (a distance considerably smaller than the diameter of a single foveal cone) suggests that considerable neural processing is involved in these hyperacuity tasks. Furthermore, interference with hyperacuity response from flanking lines has been shown to be centrally mediated [4].

In order to understand better the nature of the hyperacuity response, and to provide the basis for development of new clinical tests, we have replaced the increment threshold target of the standard sensitization paradigm with a hyperacuity target. This allows us to map out the spatial interactions about this unique type of visual stimulus in the same way as we have previously mapped the spatial interactions about the increment threshold stimulus in the standard paradigm.

In the present report, we demonstrate in normal observers an effect on foveal hyperacuity performance caused by different-sized backgrounds and map these spatial interactions in two dimensions. After establishing standards for this type of spatial interaction, we plan to apply this test in cases of visual anomaly with the goals of better understanding the normal mediation of the hyperacuity response and of establishing its clinical diagnostic ability.

Methods

All stimulus patterns were composed of white lines generated on a CRT screen under computer control. At the viewing distance of 5.7 meters, individual lines could be positioned with a resolution of 6 seconds of arc. Hyperacuity thresholds were estimated using a two-alternative forced choice method in which observers had to judge whether the upper of two small points of light spaced 4 min arc apart vertically was positioned to the left or to the right of the lower dot. The centrally fixated target was presented for .5 seconds every 3 seconds with the direction of displacement of the upper dot relative to the lower one randomized from trial to trial. Observers were given immediate feedback as to the outcome of each trial. Each stimulus dot of the two-dot target subtended 1 minute of arc and had a luminance of 490 cd/m^2. This stimulus pattern was superimposed and centered upon a variable-sized rectangular background which had a luminance of 31 cd/m^2. Within a session, either the width or the height of the background was held constant, while the other dimension was varied, with presentation order counterbalanced between sessions. Two sessions provided thresholds at nine background sizes, with each threshold determined from a probit analysis of 240 responses.

Results and Discussion

The results indicate that hyperacuity performance is differentially influenced by different widths of rectangular backgrounds of light. Widening a rectangular background indeed first drives threshold up, then down, much as widening a rectangular background alters the increment threshold of a line [5,6]. This basic finding of the present study is shown in Fig. 1 for two subjects. On this 12 minute high background, maximum threshold for both subjects is reached at a width of 3 to 4 min arc, where hyperacuity thresholds are elevated by a factor of three compared to the thresholds at the largest backgrounds where the curves are flat. Hyperacuity thresholds with no background present matched thresholds at this flat portion of the curves. These curves are generally similar to the line target and spot target increment threshold curves except that the peak increment thresholds occur at a wider background width (about 6 min arc) than the peak hyperacuity thresholds (3 to 4 min arc).

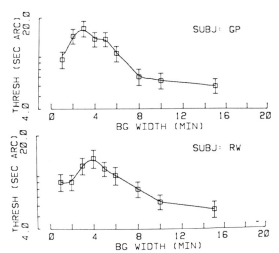

Fig. 1 Hyperacuity thresholds (in seconds of arc) of two subjects for a two-dot vernier target centered on a 12 min arc high, variable-width, rectangular background are shown as a function of the width of the background. Each threshold is based on a total of 240 responses and corresponds to the displacement from vertical alignment, of the upper dot relative to the lower dot, which is detectable 75% of the time. Error bars indicate ±1 standard error in this and all other figures.

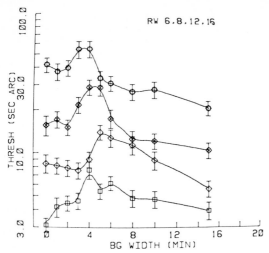

Fig. 2 Displacement thresholds of subject RW are shown as a function of background width for four different background heights, 6 min arc (open squares), 8 min arc (open diamonds), 12 min arc (filled diamonds), and 16 min arc (open circles). The 8 min arc curve is plotted at the correct absolute level; the 6, 12, and 16 min arc curves are shifted by -0.3, 0.3, and 0.6 log units, respectively, for clarity of presentation.

Figure 2 shows that differential background effects are not limited to a background height of 12 min arc. The same general pattern of threshold increase, decrease, and leveling was obtained at background heights of 6, 8, 12, and 16 min arc for both subjects tested. Peak threshold elevation occurs for backgrounds between 3 and 5 min arc wide at all background heights.

The influence of background height was examined directly by fixing the width of the background and varying its height. Two background widths were used: one which had produced the maximum threshold elevation previously (3 minutes), and one which had minimal influence on threshold (12 min arc). Varying background height at both of these widths systematically alters hyperacuity threshold. The four subjects tested under these conditions gave very similar results. Figure 3 shows individual results for two observers, and Fig. 4 shows the group means for the four observers combined. The group and the individual data show gradually increasing thresholds as the height of a 3 min arc wide background is increased (dashed curve) and gradually decreasing thresholds as the height of a 12 min arc wide background is increased (solid curve). This general finding fits with the data from the variable-width series (Figs. 1 and 2) as 3 min arc wide backgrounds were seen to raise threshold and 12 min arc wide backgrounds were seen subsequently to lower threshold. In addition, the variable-height results shown in Figs. 3 and 4 seem to incidate that the area which produces the largest hyperacuity threshold elevation is relatively narrow in width and extended in height.

The effects of varying the height of the background are directly compared to the effects of varying its width in the next two figures. Fig. 5 shows individual data for two subjects, and Fig. 6 shows group means for the four subjects combined. The abscissae now represent the extent of the variable dimension, either the varying width of a 12 min arc high background (dashed curves) or the varying height of a 12 min arc wide background (solid curves).

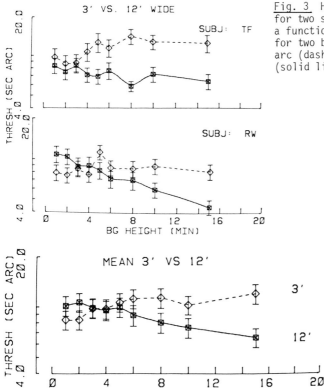

Fig. 3 Hyperacuity thresholds for two subjects are shown as a function of background height for two background widths, 3 min arc (dashed line) and 12 min arc (solid line)

Fig. 4 A comparison of group data obtained with 3 min arc wide (dashed line) and 12 min arc wide (solid line) rectangles, as a function of background height, is illustrated. Mean thresholds and mean standard errors for all four subjects tested are plotted.

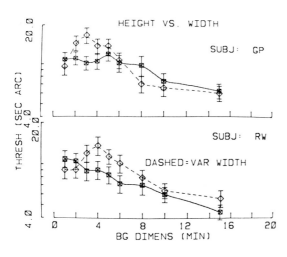

Fig. 5 The effect of varying the width of a 12 min arc high background (dashed line) is compared to the effect of varying the height of a 12 min arc wide background (solid line) for two subjects. Abscissa values indicate the extent of the varied background dimension (either height or width); see text

235

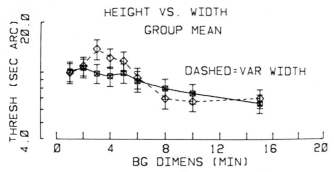

Fig. 6 Group mean thresholds (four subjects) obtained on a rectangular background varied in width (dashed line) or height (solid line), with the other dimension fixed at 12 min arc, are shown. (Details as in Fig. 5.)

Since the extents of the fixed dimensions are equal (12 min arc), the two conditions equate background area and shape (rotated 90°) at each abscissa value. All subjects show evidence of a stronger threshold elevation effect of the background when its width is expanded to either side of the hyperacuity target than when its height is increased above and below the target by a corresponding amount. Furthermore, expanding the height of a 12 min arc wide background causes a steady gradual decline of threshold, whereas expanding the width of a 12 min arc high background (i.e., rotating the same background shapes 90°) causes first a threshold elevation and then a threshold lowering. Thus, the threshold-raising effects of the background are not radially symmetrical about the hyperacuity target. Taken together, these data show that hyperacuity threshold is affected in opposite ways at near and far lateral distances, but these antagonistic areas each change monotonically in vertical extent through 15 min arc.

An alternative explanation of these hyperacuity results is suggested by the increment threshold results of the traditional sensitization paradigm. In that paradigm, varying the size of a background, such as we used here, first decreases, then increases the visibility of the central target. The results in Figs. 7 and 8 show that a corresponding influence on the perceived

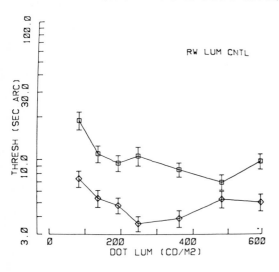

Fig. 7 Hyperacuity thresholds of subject RW as a function of the luminance of the two-dot target superimposed upon standard luminance (31 cd/m^2) backgrounds of 3 min arc width (squares) and 15 min arc width (diamonds). Background height is fixed at 12 min arc. Two-dot vernier test luminance in all other aspects of this study was 490 cd/m^2

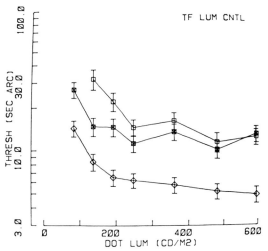

Fig. 8 Hyperacuity thresholds of subject TF as a function of the luminance of the two-dot target. Details as in Fig. 7. Data for background widths of 1 min arc (filled squares), 3 min arc (open squares), and 15 min arc (diamonds) are shown.

brightness of the central hyperacuity target cannot account for the similar pattern of hyperacuity threshold changes reported here. Hyperacuity thresholds were measured as a function of the luminance of the two-dot target (over a range of 75 to 600 cd/m²) on both narrow and wide backgrounds, all 12 min arc high. Although thresholds generally decrease with increases in luminance, this effect is large only at very low target luminances. For luminances of 250 cd/m² to 600 cd/m² the threshold curves of both subjects tested are quite flat. Thus, in the neighborhood of the target luminance employed in all other conditions of this study (490 cd/m²), even large shifts of the brightness of the target cannot account for the observed changes in hyperacuity threshold on different size backgrounds. Figure 8 also includes data for a 1 min arc wide background and illustrates that the basic pattern of a higher threshold at 3 min arc wide backgrounds and lower thresholds at both smaller (1 min arc) and larger (15 min arc) backgrounds holds across a very wide range of target luminances.

Another issue to be considered is whether the elevated hyperacuity thresholds are due to the presence of light at different positions around the hyperacuity target as in the increment threshold paradigms [3] or due instead to the presence of the edges of the rectangular background at the different distances from the hyperacuity target. WESTHEIMER and co-workers [4,7] have demonstrated a similar threshold-elevating effect of bright lines at variable distances from another type of hyperacuity target. However, since the fine lines used in those studies also contain both edge and luminance components, the relative contribution of each to the observed effects is still not clear. We are now beginning experiments to separate the two possible contributors.

Conclusions

These findings provide the basis for an examination of the effects of visual anomaly on the standard pattern of spatial interactions near a hyperacuity target. The nature of the hyperacuity response is such that it probably requires considerable central visual processing. The manner and degree in which well-characterized functional disruptions of various stages of the neural visual pathway alter these established hyperacuity spatial interactions should provide valuable information to clinical and basic visual science.

Acknowledgments

This research was supported in part by NEI Grant EY 03669, N.I.H., Bethesda, Maryland.

References

1   J.M. Enoch: Invest. Ophthal. Vis. Sci. 17, 208 (1978)
2   G. Westheimer: J. Physiol. (Lond) 181, 881 (1965)
3   G. Westheimer: J. Physiol. (Lond) 190, 139 (1967)
4   G. Westheimer and G. Hauske: Vision Res. 15, 1137 (1975)
5   K. Fuld: Vision Res. 18, 1045 (1978)
6   E.A. Essock: Stimulus orientation biases of the human visual system:
    a unifying model. Unpublished doctoral dissertation, Brown University,
    Providence, RI (1981)
7   G. Westheimer, K. Shimamura, and S.P. McKee: J. Optic. Soc. Am. 66, 332
    (1976)

Assessment of Temporal Resolution in Multiple Sclerosis by Multi-Flash Campimetry

Charles W. White, Edward M. Brussell, Olga Overbury, and Pardo Mustillo
Concordia University, Montreal, Quebec, Canada H4B 1R6

Introduction

There is good evidence that some of the major effects of multiple sclerosis
(MS) involve deficits in the temporal responses of the visual system. Naturally, several different psychophysical and electrophysiological methods have
been developed for assessing temporal sensitivity in visual disorders. Such
tests include, for example, tests of delayed visual perception such as apparent simultaneity matches [1], critical fusion frequency for flashing lights
[2,3,4], Pulfrich effect tests [5], and measurements of the latencies of
visual evoked potentials [6,7,8]. We believe that one of the most useful
measures of temporal resolution is the double-flash discrimination test that
has been developed by REGAN and others [9,10,11].

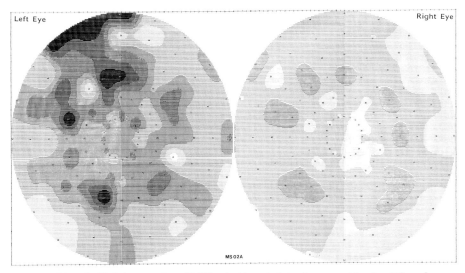

Fig. 1 Visual field map of multiflash thresholds for a patient with multiple sclerosis. The darker the map, the worse is the temporal resolution
in the corresponding region of the patient's visual field. In all the visual
field maps in this report, the data from the observer's left eye are shown
on the left, and the right eye data are on the right. The circular field
for each eye subtends a visual angle of 40 deg, and the coordinate system
is in visual field, rather than retinal, coordinates. Also, the central area
of each field has been expanded by logarithmic scaling of the radial distance
from the center of each field.

239

The results that we are presenting are derived from a modification and extension of the double-flash method for assessing temporal resolution in multiple sclerosis [12]. Its advantages include its theoretical rationale, its ease of application, and its spatial specificity. Figure 1 illustrates some of the results for one of the MS patients. It is a visual field density map in which the dark areas represent regions in which the temporal sensitivity is impaired. Notice that the right eye is better, and the left eye is worse, with several discrete islands of temporal impairment. The map was generated by a computer mapping program [13] that interpolates among the 240 independent measurements of temporal resolution.

Method

The traditional double-flash method is a temporal two-pulse threshold in which psychophysical methods are used to determine the minimum interval that allows two brief flashes to be discriminated from a single flash of comparable energy. The stimuli to be discriminated are two pulses of light separated by a variable interval. If the interflash interval is very short, less than 20 msec, for example, then the two flashes are not resolved and the observer reports seeing only a single flash. However, if the interval is increased, to 200 msec, for example, then the observer reports seeing two clearly defined flashes. The two-flash threshold is defined as the minimum interval which permits the two flashes to be resolved reliably.

We have extended this technique by presenting a series of flashes, rather than just two, and by continuously increasing the amount of time separating the flashes until the observer reports seeing a flickering point of light rather than a steadily illuminated point. Another way to describe this manipulation is that of decreasing the duty cycle (i.e., reducing the proportion of on-time) until the light appears to flicker. Observers ordinarily report

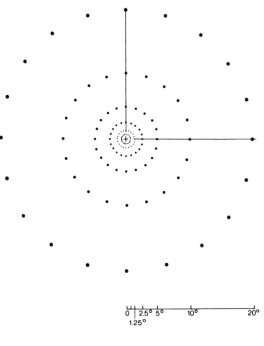

Fig. 2 Multiflash stimulus display. The 120 test points are arranged in six concentric rings. The outer ring subtends 40 deg of visual angle, and each other ring has one-half the diameter of the next larger ring. During testing, the four quadrants of the display were presented in random order. The individual points are not drawn to scale, and the points in the center have been reduced for illustrative purposes (from [12])

seeing a steadily illuminated point when the off-period is short and, under our viewing conditions, normal observers usually report first seeing flicker whenever the off period exceeds approximately 25 msec.

The stimulus display, as illustrated in Fig. 2, consists of 120 points arranged in six concentric circles. The radial separation of the 20 points in each circle is 18 polar degrees. At the viewing distance of 57 cm, each point subtends a visual angle of about 5 min of arc. The diameter of the outermost circle subtends 40 deg of visual angle. Since the radius of each smaller circle is one-half that of the previous one, the radius of the inner-most circle subtends a visual angle of only 1.25 deg. During testing, the origin of the display coordinate system is shifted to the corners of the CRT screen, so that only one quadrant of the display is visible at any given time. The computer presents each of the quadrants to each eye in a random order, and the order in which the points are tested within each quadrant is also randomly determined. When the temporal sensitivity of a particular retinal location is to be assessed, the computer flashes the appropriate point at a frequency of 5 Hz. This basic frequency was chosen because it is one to which most people are highly sensitive, and it is fast enough to allow a duty cycle threshold to be reached in a few seconds. Within each 200 msec cycle, the duty cycle was decreased in steps of 1.1 or 2.8 msec until the observer pressed a key to indicate the appearance of a flickering point.

Procedure

In a preliminary practice period, the observer was given an opportunity to practice the task on a simplified display of only 12 points, by fixating a small cross-shaped target and pressing a response key as soon as any one of the twelve displayed points appeared to flicker. One of the 12 practice points was positioned such that its image fell on the optic disk, in the blind spot, as long as proper fixation was maintained. Under these condi-tions, that point should never have been seen, so that a flicker response to that point indicated improper fixation. This provided an early opportun-ity to detect observers who were unable to fixate properly, as well as a check on the comprehension of the instructions. When we were convinced that the observer was capable of taking the test, and the observer felt comfort-able with the test conditions, a fixation cross was presented in one corner of the screen followed by the first quadrant of the display.

At the end of each session the computer printed the critical off-periods for each quadrant and for both eyes (120 points per eye), and noted any points that were statistically deviant, either from other points at the same retinal eccentricity, or from the overall average value for each eye. The statis-tically deviant points were then immediately repeated in order to ascertain whether they reflected lapses in attention, or momentary loss of fixation, or whether they corresponded to retinal regions whose temporal resolving power was genuinely impaired.

Twenty control subjects were tested, none of whom suffered from any known ophthalmological disorders, and all of whom had 20/20 acuity or were correc-ted to 20/20. Seventeen MS patients were tested, whose conditions varied from no immediately observable symptoms to almost complete confinement to a wheelchair. All but one MS patient had a history of optic neuritis and no patient suffered from any other ophthalmological disorder. The corrected visual acuities of the MS patients were at least 20/20 with the exception that one patient had a large central scotoma in one eye, which had developed over a period of years. The control subjects ranged in age from 21 to 39 with a mean age of 28 years, and the patients ranged in age from 20 to 63

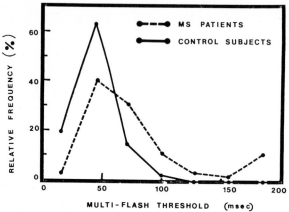

Fig. 3 Relative frequency distributions of multiflash thresholds (critical off-periods) for MS patients and control observers.

with a mean of 45 years. All subjects wore their corrective lenses during the testing sessions.

Results and Discussion

The graph in Fig. 3 represents an overall summary of the results. The multi-flash thresholds (critical off-periods) were classified into seven intervals between zero and 200 msec and converted to relative frequency distributions. The MS patients have relatively fewer thresholds below approximately 60 msec, and relatively more above that point. The MS patients also exhibit a secon-dary peak above 150 msec. Thus, there is a clear differentiation between the MS patients and the control observers, even at this gross level of analy-sis. The differences in Fig. 3 are not as striking as they might be, however, because the data are averaged over large areas of the visual fields, which tends to minimize the local discontinuities in the MS patient data.

In order to illustrate the differences between the MS patients and the control observers, it is necessary to consider the spatial configuration of the data from individual observers. Fortunately, the multiple flash tech-nique samples so many retinal locations that it is possible to create exten-sive maps of the temporal resolution in the visual field. Some of these are illustrated in Fig. 4. Each plot represents a 40 deg circular visual field for each eye. The retinal eccentricity, i.e., the radial distance from the center of each field, is represented on an octave scale. In other words, the radii of the concentric rings of test points have been scaled so that the rings are equally spaced. This produces a relative magnification of the central region of the visual field which corresponds to its greater func-tional significance, and enhances the spatial resolution of the test area. The 200-msec range of possible critical off-periods is divided into a 7 cate-gory gray scale such that each category reflects a 28.5 msec range of criti-cal off-periods. The darker the shading in these visual field maps, the worse is the temporal resolution.

In general, the visual field maps for the control observers, in the left-hand column, provide a characteristic picture of good temporal resolution in the central visual field with a slight tendency towards somewhat worse resolution (darker regions) in more peripheral areas. The visual field maps

242

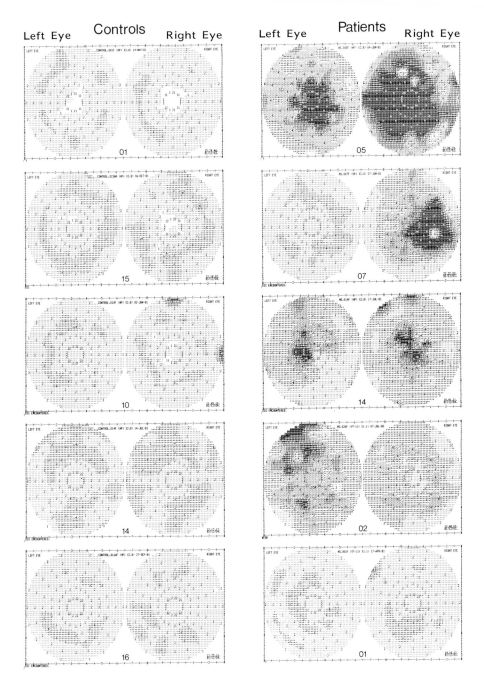

Fig. 4 Visual field maps for five MS patients and five control observers. The dark "islands" in the patients' data represent local regions of impaired temporal sensitivity (from [12]).

for the MS patients, in the right-hand column, however, demonstrate a wide range of anomalous temporal sensitivity distributions. The MS patient maps in Fig. 4 are arranged with the severity of loss in temporal resolution decreasing down the column. At the upper end of this continuum, there are patients with severe, large bilateral islands of impaired temporal resolution. Other patients have smaller islands for which the difference in temporal resolving power between them and other regions is less exaggerated. It is these islands of varying severity that usually distinguish the MS patients from the control observers. By this criterion there is only one patient we have tested whose visual field maps could not be easily discriminated visually from the control observer maps. That patient's data are plotted at the bottom of the right-hand column in Fig. 4.

The fact that the patient data are characterized by islands answers a possible criticism of our current results. The data that we have reported might suffer from the fact that our patient sample was older, on the average, than our control sample. To what extent can the differences between the data collected from the two groups be accounted for by the age differences? Although CFF and the double-flash threshold have been shown to deteriorate with age, there is no evidence or logical reason to suspect that these measures, or temporal resolution in general, deteriorates in anything but a uniform fashion across the visual field. That is, we would expect a general decrease in temporal resolution in older control observers, but not the occurrence of the discrete islands that are so prevalent in the MS patient data. In any event, the best strategy to settle that question would be to establish normative data for different age groups in additional experiments.

One of the most important considerations for any psychophysical procedure that is used in diagnosis or assessment is the reliability of the test, that is, how well do the test data replicate on subsequent retesting? In order to be able to evaluate the retest reliability of the multiflash procedure, we have retested several of the MS patients from the original sample, and the test-retest data for one MS patient are depicted in Fig. 5.

The upper panel in Fig. 5 does not exhibit many serious islands of temporal impairment, except for a few points to the extreme left in the left eye field and to the right in the right eye field. Notice, however, that there are two minor impaired areas in the left eye field, located at approximately 10 o'clock and 4 o'clock, at a retinal eccentricity of 5 deg. Those areas take on additional significance when they are compared to the lower map in Fig. 5, which represents data from a retesting session approximately eight months after the first test. The small islands in the left eye field are enlarged and darker in the retest map. On careful examination, similar effects also appear for several areas in the right eye as well. An accurate evaluation of the retest reliability will require additional testing, of course. Nevertheless, the data illustrated in Fig. 5 strongly suggest that the multiflash procedure can produce replicable results. It also indicates that the technique may prove capable of monitoring subtle changes in temporal sensitivity in progressive disorders.

Conclusions

We have reported an extension of the standard double-flash discrimination test that allows the temporal resolution of 240 retinal locations to be assessed in about half an hour. Given the large number of retinal locations that can be quickly sampled, the data allow for the creation of visual field maps that can be used to distinguish visually between patients and normals. With the increasing power of relatively inexpensive microprocessing systems, multiflash campimetry may provide another useful tool for the neuro-ophthalmologist in particular, and for ophthalmologists in general.

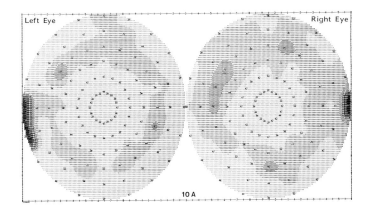

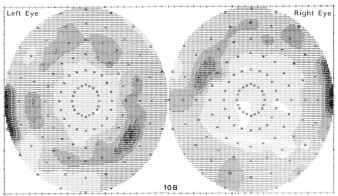

Fig. 5 Test-retest pair of visual field maps for a patient with MS. The follow-up data, shown in the lower map, were obtained approximately 8 months after the original test session, shown in the upper map.

Acknowledgments

We appreciate the assistance of the Lethbridge Rehabilitation Centre, the Lakeshore chapter of the Multiple Sclerosis Society of Canada, and especially Allan Putterman, Rachel Gunner, and Wilma Mallick for their cooperation in referring patients to us. We also thank Terrence Williams and Stanley Ròg, for their assistance in preparing the figures. This project was partially supported by the Natural Sciences and Engineering Research Council of Canada and by Le Programme de Formation de Chercheurs et d'action Concertée (FCAC) du Québec.

References

1 J.R. Heron, D. Regan, and B.A. Milner: Brain 97, 69 (1974)
2 O.A. Parsons and P.N. Miller: Arch. Neurol. 77, 323 (1957)
3 A.F. Titcombe and B.G. Willison: J. Neurolog. and Neurosurg. Psychiat. 24, 260 (1961)
4 M.L. Daley, R.L. Swank, and C.M. Ellison: Arch. Neurol. 36, 292 (1979)

5   D. Rushton: Brain 98, 283 (1975)
6   A.M. Halliday, W.I. McDonald, and J. Mushin: Lancet 1, 982 (1972)
7   A.M. Halliday, W.I. McDonald, and J. Mushin: Brit. Med. J. 4, 661 (1973)
8   I. Bodis-Wollner, C.D. Hendley, L.H. Mylin, and J. Thornton: Ann. Neurol.
    5, 40 (1979)
9   R.J. Galvin, D. Regan, and J.R. Heron: Brain 99, 255 (1976)
10 D. Regan: New visual tests in multiple sclerosis. In *Topics in Neuro-
    Ophthalmology*, ed. by H.S. Thompson, R. Daroff, L. Frisen, J.S. Glaser,
    and M.D. Sanders (Williams & Wilkins, Baltimore, 1979)
11 D. Regan: Detection and quantification of neuro-ophthalmological abnor-
    malities using psychophysical measures of visual delay and temporal
    resolution. In *Electrophysiology and psychophysics: Their use in ophthal-
    mic diagnosis*, ed. by S. Sokol (Little, Brown & Company, Boston, 1980)
12 E.M. Brussell, C.W. White, M. Bross, P. Mustillo, and M. Borenstein: Cur-
    rent Eye Res. 1, 671 (1981/82)
13 J.A. Dougenik and D.E. Sheehan: *SYMAP User's Reference Manual* (Harvard
    University Laboratory for Computer Graphics and Spatial Analysis, Cam-
    bridge, Mass., 1975)

Index of Contributors

Springer Series in Optical Sciences

Editorial Board: **J.M.Enoch, D.L.McAdam, A.L.Schawlow, K.Shimoda, T.Tamir**

Volume 23
Vertebrate Photoreceptor Optics

Editors: **J.M.Enoch, F.L.Tobey, Jr.**
With a Foreword by W.S.Stiles
1981. 164 figures. XV, 483 pages
ISBN 3-540-10515-8

Contents: *J.M.Enoch:* Introduction. – *B.Borwein:* The Retinal Receptor: A Description. – *J.M.Enoch:* The Stiles-Crawford Effects. – *J.M.Enoch:* Retinal Receptor Orientation and Photoreceptor Optics. – *J.M.Enoch:* Waveguide Properties of Retinal Receptors: Techniques and Observations. – *B.R.Horowitz:* Theoretical Considerations of the Retinal Receptor as a Waveguide. – *W.Wijngaard:* Theoretical Consideration of Optical Interactions in an Array of Retinal Receptors. – *R.Winston:* The Visual Receptor as a Light Collector. – *F.I.Harosi:* Microspectrophotometry and Optical Phenomena: Birefrigence, Dichroism, and Anomalous Dispersion. – *J.A.C.Micol:* Tapeta Lucida of Vertebrates. – *G.D.Bernard:* A Comparison of Vertebrate and Invertebrate Photoreceptors. – Additional References with Titles. – Subject Index.

Volume 22
Lasers in Photomedicine and Photobiology

Proceedings of the European Physical Society, Quantum Electronics Division, Conference, Florence, Italy, September 3–6, 1979
Editors: **R.Pratesi, C.A.Sacchi**
1980. 108 figures, 20 tables. XIII, 235 pages
ISBN 3-540-10178-0

Contents: General Introduction to Photomedicine and Photobiology. – Photodynamical Therapy of Tumors. – Photodermatology. – Phototherapy of Hyperbilirubinemia. – Absorption and Fluorescence Spectroscopy. – Raman and Picosecond Spectroscopy.

Volume 18
Holography in Medicine and Biology

Proceedings of the International Workshop, Münster, Federal Republic of Germany, March 14–15, 1979
Editor: **G.v.Bally**
1979. 224 figures, 2 tables. IX, 269 pages
ISBN 3-540-09793-7

Contents: Introductory Survey. – Holography in Orthopedics. – Moiré Topography. – Holography in Biology. – Holography in Radiology. – Holography in Ophthalmology. – Holography in Urology. – Holography in Dentistry. – Holograpy in Otology. – Acoustical Holography. – Special Holographic Techniques. – Index of Contributors.

Volume 8
Frontiers in Visual Science

Proceedings of the University of Houston College of Optometry Dedication Symposium, Houston, Texas, USA, March, 1977
Editors: **S.J.Cool, E.L.Smith III**
1978. 533 figures, 28 tables. XIV, 798 pages
ISBN 3-540-09185-8

Contents: Keynote Presentation. – Ocular Physiology and Pathology. – Contact Lenses. – Color Vision. – Spatial Vision and Form Vision. – Binocular Vision and Stereopsis. – Neurophysiology of Visual System Function. – Development of Visual System Function. – Vision Health Care Delivery.

Springer-Verlag
Berlin
Heidelberg
New York
Tokyo

Volume 31
Optics in Biomedical Sciences
Proceedings of the International Conference,
Graz, Austria, September 7–11, 1981
Editors: **G.v.Bally, P.Greguss**

1982. 212 figures. X, 274 pages
ISBN 3-540-11666-4

Contents: Unconventional Imaging in Microscopy. – Image Processing. – Interferometry
and Holography. – Speckle-Techniques and
Spectroscopy. – Optometry. – Moiré
Methods. – Closing Remarks. – Index of Contributors.

Volume 27
D.L.MacAdam
Color Measurement
Theme and Variations
1981. 92 figures, 4 colorplates. XIII, 229 pages
ISBN 3-540-10773-8

Contents: The Physical Basis of Color Specifi-
cation. – Sources of Light. – Spectrophoto-
metry. – Color Mixture. – Determination of
Tristimulus Values. – Color of Light. – Colors
of Objects. – Color Differences. – Color-Order
Systems. – Color-Matching Functions. –
Chromatic Adaptation. – Notes and Sources.
– Subject Index. – Author Index.

Volume 19
G.A.Agoston
Color Theory and
Its Application in Art
and Design
1979. 55 figures, 6 color plates, 12 tables.
XI, 137 pages
ISBN 3-540-09654-X

Contents: Introduction. – Color: Two Con-
cepts. – Perceived Colors. – Light and Color.
– Colored Materials. – Color Specification
(CIE). – Diverse Applications of the CIE
Chromaticity Diagram. – Color Systems. –
Appendix. – Color Plates I-VI. – References.
– Author and Subject Index.

Volume 13
Y.Le Grand, S.G.EL Hage
Physiological Optics
1980. 118 figures, 9 tables. XVII, 338 pages
ISBN 3-540-09919-0

Contents: Introduction. – Description of the
Human Eye. – The Dimensions of the Eye. –
Optics of the Eye. – The Retinal Image. –
Accommodation. – Theories of Accommoda-
tion. – Spherical Ametropia. – Correction of
Ametropia. – Spherical Corrective Lenses. –
Corneal Astigmatism. – Vision of the Astig-
matic. – The Orbit and Its Muscles. – Geo-
metry of the Movements of the Eye. – Cor-
rection of the Mobile Eye (Far Vision). – Cor-
rection of the Mobile Eye (Near Vision-Astig-
matic Eye). – Normal Binocular Movements.
– Anomalies of Binocular Movements. – Cor-
rective Lenses and Binocular Vision. –
Measurement of Visual Acuity. – Subjective
Optometers. – Ophthalmoscopes. – Objective
Measurement of Refraction. – Exercises. –
Solutions to the Exercises. – References. –
Author Index. – Subject Index.

Springer-Verlag
Berlin
Heidelberg
New York
Tokyo